Supersonic Business Jets in Preliminary Aircraft Design

TU Braunschweig – Niedersächsisches Forschungszentrum für Luftfahrt

Berichte aus der Luft- und Raumfahrttechnik

Forschungsbericht 2016-04

# Supersonic Business Jets in Preliminary Aircraft Design

**Martin Schuermann**

TU Braunschweig

Institut für Flugzeugbau und Leichtbau

---

Diese Arbeit erscheint gleichzeitig als von der Fakultät für Maschinenbau der Technischen Universität Carolo-Wilhelmina zu Braunschweig zur Erlangung des akademischen Grades eines Doktor-Ingenieurs genehmigte Dissertation.

Bibliografische Information der Deutschen Nationalbibliothek
Die Deutsche Nationalbibliothek verzeichnet diese Publikation in der Deutschen Nationalbibliografie; detaillierte bibliografische Daten sind im Internet über http://dnb.d-nb.de abrufbar.
1. Aufl. - Göttingen: Cuvillier, 2016
Zugl.: (TU) Braunschweig, Univ., Diss., 2016

Diese Arbeit erscheint gleichzeitig als von der Fakultät für Maschinenbau der Technischen Universität Carolo-Wilhelmina zu Braunschweig zur Erlangung des akademischen Grades eines Doktor-Ingenieurs genehmigte Dissertation.

**Herausgeber der NFL Forschungsberichte:**
TU Braunschweig – Niedersächsisches Forschungszentrum für Luftfahrt
Hermann-Blenk-Straße 27 • 38108 Braunschweig
Tel: 0531-391-9822 • Fax: 0531-391-9804
Mail: nfl@tu-braunschweig.de
Internet: www.tu-braunschweig.de/nfl


Nonnenstieg 8, 37075 Göttingen
Telefon: 0551-54724-0
Telefax: 0551-54724-21
www.cuvillier.de


1. Auflage, 2016
Gedruckt auf umweltfreundlichem, säurefreiem Papier aus nachhaltiger Forstwirtschaft.

ISBN 978-3-7369-9278-8
eISBN 978-3-7369-8278-9

# Supersonic Business Jets in Preliminary Aircraft Design

Von der Fakultät für Maschinenbau der
Technischen Universität Carolo-Wilhelmina zu Braunschweig

zur Erlangung der Würde eines
Doktor-Ingenieurs
(Dr.-Ing.)

genehmigte

Dissertation

von

Dipl.-Ing. Martin Schuermann

aus Celle



| | |
|---|---|
| eingereicht am: | 18. September 2015 |
| mündliche Prüfung am: | 11. Mai 2016 |
| Referenten: | Prof. Dr.-Ing. Peter Horst |
| | Prof. Dr.-Ing. Volker Gollnick |
| Vorsitzender: | Prof. Dr.-Ing. Peter Eilts |

2016

for my family

# Acknowledgements

The author owes a substantial debt of gratitude to many individuals, colleagues and family members, who have contributed thoughts or inspirations used in this dissertation. Although there are too many individuals to list exhaustively, several ones of them must be acknowledged for their singular importance. This includes first and foremost Prof. Peter Horst, who employed me at the Institut für Flugzeugbau und Leichtbau of the TU Braunschweig and provided support and advice to me during the preparation of this thesis. Moreover, I would like to express my sincere thanks to Prof. Gollnick and Prof. Eilts for their involvement in the examination committee. Part of this thesis is based on work performed within the VIP-3 project of the German Aerospace Centre (DLR). Bernd Liebhard and Michele Gaffuri worked with me in this project and I do thank them for their advice and fine cooperation. I likewise would like to thank Dr. Florian Hürlimann who provided a verification model which became an indispensable asset. My thanks go to my colleagues at IFL for many productive and enjoyable moments. Moreover, I owe special thanks to my officemate Tayson Weiss and to Paul Reich. Their inspiring and critical remarks lead to a lot of fruitful discussions.
Finally, I am indebted to my family who supported me in the pursuit of this work and especially to my wife for her unconditional support and patience.

# Abstract

The Preliminary Aircraft Design and Optimisation tool, PrADO, is an in-house program of the Institute of Aircraft Design and Lightweight Structures, TU Braunschweig, Germany, which covers a wide range of aspects of aircraft preliminary design. An initial aircraft concept serves as a basis for various analysis modules. *"Each module is designated to fulfil one special task e.g. aerodynamic analysis, estimation of structural mass, etc. The available methods grouped within those modules range from statistical methods to physics based models"* (Schuermann et al., 2015). From an aircraft developer's point of view PrADO is used within both, the conceptual and the preliminary design phase.

The aim of this thesis is to introduce methods and methodologies to aircraft conceptual and preliminary design, more precisely to PrADO, that allow to judge supersonic aircraft concepts. Therefore, the aerodynamic analysis module, the structural analysis module and the propulsion module are extended. An inviscid flow solver is integrated to obtain aerodynamic coefficients. The calculated data serves as input to other analysis modules of PrADO. While the aerodynamic analysis module solely uses the outer geometry of the aircraft, the structural analysis module uses its internal structural layout as additional input to a herein developed finite element model generator. The distribution of secondary mass, fuel loading and payload distributions as well as loads for ground cases and trimmed flight cases are taken from the PrADO database, whereas the aerodynamic forces are calculated by solving the inviscid Euler equations. The model serves as basis for structural sizing and consequently the estimation of structural mass. The purpose of the propulsion module is to size the engine, to calculate the engine performance map and to provide reliable mass data based on the thermodynamic cycle. PrADO provides various models for the analysis of turbojet, turbofan and turboprop engines. It is extended by a turbofan engine with mixed exhaust flow.
The verification of the aerodynamic data is solely based on its expected qualitative distribution, since data from higher order methods was not available for the time being. The results compare well with the expected behaviour. The structural sizing process is verified based on an example from literature. The results of the developed sizing algorithm show extraordinary agreement with the reference data. The extended aircraft design process is finally applied to an aircraft from the European research project HISAC. The results compare well with global aircraft data.
An excursion into the field of temperature effects in supersonic flight is provided, since no relevant literature is found on the topic with regard to conceptual and preliminary aircraft design. The results are translated into helpful information on the material selection process in the stage of aircraft pre-design.
Eventually, the tool chain is applied to analyse a supersonic business jet and the results are presented. Based on the results of this basic design, a parameter study is conducted. A combination of cruise Mach number and design range is varied. Global design parameters show expected sensitivity to such variations.

# Kurzfassung

Das am Institut für Flugzeugbau und Leichtbau der TU Braunschweig entwickelte Programm PrADO (Preliminary Aircraft Design and Optimisation tool) stellt Werkzeuge zur Analyse und Beurteilung von Flugzeugen in der Konzept- und Vorentwurfsphase bereit. Basis der Analyse ist ein erstes Flugzeugkonzept. Jedes Entwurfsmodul gruppiert Methoden zur Lösung einer Teilaufgabe im Vorentwurfsprozess, z.B. Bestimmung der aerodynamischen Beiwerte oder Masseanalyse. Die vorhandenen Methoden reichen von einfachen Abschätzungsformeln bis hin zu physikalisch begründeten Berechnungsmodellen.

Die Zielsetzung dieser Arbeit ist es, Methoden in die Konzept- und Vorentwurfsphase, genauer in das Programm PrADO, einzuführen, welche die Bewertung von Überschallflugzeugen ermöglichen. Daher werden die aerodynamischen, strukturmechanischen und antriebsspezifischen Entwurfsmodule erweitert. Zur Berechnung der aerodynamischen Beiwerte wird ein Euler-Verfahren eingesetzt. Die Ergebnisse stehen anderen Entwurfsmodulen in Form von Kennfeldern zur Verfügung. Das strukturmechanische Modell benötigt, neben der äußeren Geometrie, auch den inneren Aufbau des Flugzeugs. Diese Daten bilden die Grundlage eines Finite-Elemente-Modells. Die Verteilung von Sekundärmassen, Kraftstoff und Nutzlast, sowie Lasten für Boden- und ausgetrimmte Flugfälle werden der PrADO-Datenbank entnommen. Die aerodynamischen Kräfte werden mit dem Euler-Verfahren berechnet und auf das Modell aufgeprägt. Das Modell bildet die Basis zur Strukturdimensionierung und somit der Berechnung der Strukturmasse. Die Aufgabe des Antriebsmoduls ist die Triebwerksgröße zu bestimmen, Triebwerkskennfelder bereitzustellen und die Masse des Antriebssystems zu berechnen. Das Modul wird um ein Zweikreis-Turbinenluftstrahltriebwerk mit Mischung von Primär- und Sekundärstrom ergänzt.

Die qualitativen Verläufe der aerodynamischen Beiwerte entsprechen dem zu erwartenden Verhalten. Für einen quantitativen Vergleich stehen keine Vergleichsdaten aus höherwertigen Verfahren zur Verfügung. Das erweiterte Strukturanalysemodul wird auf ein Literaturbeispiel angewendet. Die Ergebnisse zeigen eine sehr gute Übereinstimmung mit den Daten. Nach der Verifikation der Einzelmodule wird der Gesamtentwurfsprozess an einem Flugzeug des europäischen Projekts HISAC überprüft. Die globalen Entwurfsergebnisse zeigen sehr gute Übereinstimmungen mit den verfügbaren Referenzdaten. Die im Überschallflug auftretenden thermischen Aspekte werden in einem kurzen Exkurs behandelt. Dies erscheint notwendig, da zu diesem Thema keine, auf den Vorentwurf bezogene Literatur gefunden wurde. Die Ergebnisse geben dem Ingenieur erste Hinweise zur Materialauswahl im Rahmen der Konzept- und Vorentwurfsphase.
Schlussendlich wird der erweiterte Entwurfsprozess zum Entwurf eines Überschallgeschäftsreiseflugzeugs genutzt. Auf Basis der gewonnenen Erkenntnisse wird eine Parameterstudie durchgeführt. Die Sensitivitäten der globalen Entwurfsparameter bei Änderungen von Machzahl und Reichweite werden gut wiedergegeben.

# Contents

# List of Figures

# List of Tables

# 1 Introduction

## 1.1 Motivation

The design of aircraft is a fascinating yet challenging task. Frequently, opposing goals are to be fulfilled and limitations, often imposed by regulations, are to be met. However, the main design goals have always been safety and reliability, though in the last decades ecological and economical issues have complemented the former. Therefore, an aircraft design is always the result of careful consideration of all these aspects and consequently a, not only technical, compromise. The basic geometrical layout of aircraft has not changed much since the beginning of the $20^{\text{th}}$ century; nonetheless, its technological complexity has changed tremendously. One example is lightweight design where new possibilities to save weight have been used by introducing high performance aluminium alloys and composite materials. Another example are the advances in avionics and electrical systems design, resulting in a more and more "electric" aircraft. All these developments require judging their influence on an aircraft's design and performance at an early development stage to avoid economic misjudgement. This is where conceptual and preliminary aircraft design comes into play (cf. chapter 2).
Besides transportation at subsonic and transonic speeds, the dream of supersonic travel is appealing to many people and institutions. However along with military aircraft, only Concorde and the TU-144 have been introduced into the airliner market. Both aircraft have been used on very few routes and their commercial success was remote, being a good example to show that what is technically feasible must not always be economically sensible. Nevertheless, the thrill to "go supersonic" prevails and research effort and money are still being invested into the topic. Yet the focus shifts from airliners to supersonic business jets (SSBJ) and the niche of high-net-worth individuals. It is especially attractive for executives and VIPs because of prestige, convenience, comfort and the reduction of travel time. *"This listing does not claim to be complete; however, these parameters could result in an increase of corporate productivity, hence justifying supersonic business travel. The sonic boom, noise at take-off and landing, high fuel consumption, and resulting emissions are seen as critical issues for supersonic operations"* (Schuermann et al., 2015). Advances in engine technology and airframe design help to find adequate answers for the ecological and technical challenges that are related to supersonic flight. Since these issues are strongly related to the aircraft's size, an aircraft of the size of a business jet can be seen as a good starting point into practical supersonic flight. *"Recent market research has revealed that a significant number of premium passengers are willing to change to supersonic service"* (Schuermann et al., 2015). It has been shown that supersonic aircraft of the size of a business jet may seem to find a

market (Henne (2005); Liebhardt & Lütjens (2011); Liebhardt et al. (2011)). The book by Davies (1998) should not be left unmentioned here as a document of critical thinking on the issue.

The aim of this thesis is to introduce methods and methodologies to aircraft conceptual and preliminary design that allow to judge supersonic aircraft concepts. The aim of this thesis is NOT the design of an optimized aircraft for a given task! Therefore, the Preliminary Aircraft Design and Optimisation tool (PrADO) of the Technische Universität Braunschweig, Germany, is extended to suit this purpose. The focus lies on the introduction of an inviscid flow solver to obtain aerodynamic data and finite element analysis to provide means to assess structural weight. Thoughts and ideas on the principle of fully stressed design and on the sizing of structures built from composite materials are being developed and discussed.

## 1.2 Research Efforts and Existing Supersonic Projects

A great deal of research effort has been put into the investigation of commercial supersonic flight over the last decades, with concepts of supersonic business jets dating back into the late 1990's. *"Mavris & Hayden (1996) present results from a conceptual design study for a supersonic business jet serving the Pacific Rim. The results are obtained by applying the response surface methodology and design of experiments. The resulting cruise speed is Mach 2.0 with a cruise range of 3160 nautical miles"* (Schuermann et al., 2015). A different approach is proposed by Chudoba et al. (2008, 2009). They establish the hypothesis that the conversion of an existing airframe will lead to a reduction of production and purchase costs. The set of papers includes a market study and a feasibility study. The authors' goal is to be *"first in the market"* rather than providing a high technology aircraft. It is concluded that such approach is feasible, however Chudoba et al. (2009) state that *"the overall design suffers from operational limitations which are expected to severely reduce its market penetration potential"*. Other publications focus on issues that are generally of importance for a supersonic business jet and commercial supersonic flight. Notably these are emissions as pointed out e.g. by Grewe (2007), the noise of the sonic boom and general aircraft noise as outlined by, e.g., Simmons & Freund (2005), Aronstein & Schueler (2004) or Mack (2003).

*"Two prominent commercial projects linked to supersonic business travel are driven by the Aerion cooperation and by SAI (Supersonic Aerospace International). Both companies are based in the United States. The Aerion configuration aims at transporting up to twelve passengers in a comfortable cabin arrangement over a range greater than 4000 nautical miles (transatlantic capability) at a cruise speed of Mach 1.6. Its key technology is a Natural Laminar Flow (NLF) wing. A Pratt & Whitney JT8D-200 engine has been chosen to power the 41-ton (90000 lb) take-off weight jet"* (Schuermann et al., 2015). The aim is to comply with modern noise regula-

tions. Wind tunnel testing and in-flight testing of components have been conducted according to Aerion (2013).
The SAI design has been developed in cooperation with Lockheed Martin. It is a "low-boom design", which in this case means that the sonic overpressure is not higher than 24 Pa (0.5 psf). Extensive feasibility studies have been conducted, including in-flight testing of sonic boom mitigation methods. Range and payload characteristics are similar to that of the Aerion jet. SAI (2013) and Paulson (2013, 2007) give insight into the projects.
*"In addition, the European research project HISAC (environmentally friendly High Speed AirCraft) lead by Dassault Aviation has been launched within the sixth Framework Programme of the European Union"* (Schuermann et al., 2015). According to Stoufflet et al. (2008), the HISAC project's aim was to establish *"the technical feasibility of an environmentally friendly supersonic business jet ... "*. The project incorporated three different aircraft configurations, a "low-noise", "long-range" and "low-boom" arrangement. *"Its mission characteristics are similar to the ones mentioned above. Herrmann and Laban published two conference papers regarding the subject of multidisciplinary optimization applied to small supersonic aircraft"* (Schuermann et al., 2015). The first paper (Laban & Herrmann, 2007) focusses on the multidisciplinary framework, whereas the second paper (Herrmann & Laban, 2007) presents the results. The authors optimize a wing exposed to a single load case with results being fed back into the overall design loop. Their work is related to the HISAC project. Brezillon et al. (2011) present results with regard to low-boom considerations that are linked to the HISAC project. A good overview on the emergence of supersonic business travel is provided by Wiley (2007). Liebhardt et al. (2013) put some effort into the routing of supersonic aircraft to satisfy noise and supersonic boom regulations.

## 1.3 Thesis Outline

The concept of multidisciplinary aircraft design is outlined in Chapter 2. Chapter 3 presents the relevant theoretical aspects, providing the necessary background for the upcoming considerations and investigations. The focus is put on aerodynamics, structural mechanics and propulsion since these are the major disciplines on which work has been done and the author's ideas and suggestions are developed. The content of Chapter 4 gives insight into modelling aspects, explaining how the discussed theories are put to use. Verification examples are given in Chapter 5 and the field of temperature in supersonic flight is briefly discussed as an excursion in Chapter 6. All previous efforts are used to finally design a supersonic business jet, portrayed in Chapter 7. In Chapter 8 the thesis is summarized, conclusions are drawn and suggestions on future work are made.

# 2 Multidisciplinary Aircraft Design

The multidisciplinary aircraft design process is outlined in this chapter. Special emphasis is put on the conceptual and preliminary design phase in Section 2.1, since this is the context within which the author's work is carried out. Furthermore, an overview of existing aircraft design tools is given in Section 2.2 and the herein used Preliminary Aircraft Design and Optimisation tool (PrADO) is introduced in Section 2.3.

## 2.1 The Aircraft Design Process

The development of a product beginning with the idea, advancing further to its market entry and finally to its end of life can be subdivided into multiple phases. Figure 2.1 shows such a very simplistic scheme for an aircraft, where it should be noted that phases taking place after the "entry into service" (EIS) are not shown.

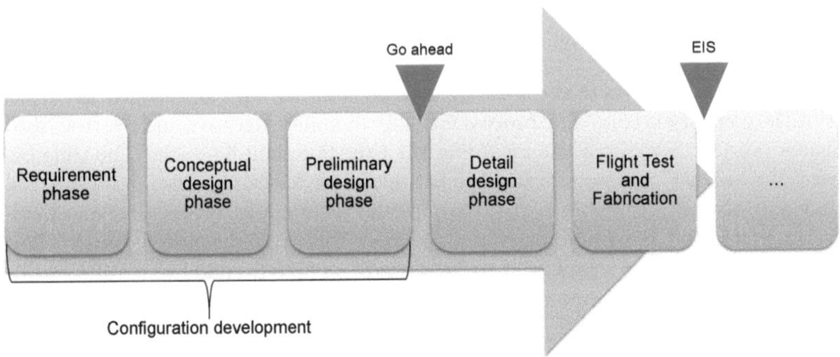


**Figure 2.1:** Basic outline of the aircraft design process

At the very beginning of the product development process, the question arises of what are the requirements to create a successful product. In the so-called *requirement phase*, data from market analysis resulting in a thorough description of customers needs is used to write a product definition. Furthermore, the requirement list is supplemented by technical constraints and regulations.
The *conceptual design phase* is initiated after successful completion of the requirement phase. This process is highly creative as solutions of any kind are considered. Thus, the impact of key technologies on the design is determined and the assessment of technological and economical risks is a vital part of the process. Analysis is usually

done by the means of statistical methods, handbook methods and existing know-how. The number of input parameters is limited to a few. The investigated designs are ranked by their compliance with the list of requirements, technical feasibility is judged and economic aspects are reviewed. Based on the results of this process, a desired solution is selected for further investigation. The outer shape of the aircraft is fixed at the end of the conceptual design phase, that is, the arrangement of its components is decided on, and key technologies are selected.

The designs being selected at the end of the conceptual design phase provide the starting point to proceed into the *preliminary design phase*. In this phase of the process, the level of detail is increased and methods of higher accuracy, often based on physics, are applied during analysis. Wind tunnel and structural tests provide input to increase accuracy. Here again, risks are assessed. At the end of the preliminary design phase, a detailed set of data for the design is available, providing a sound basis for a possible realization of the project. After completion of this phase the management will decide whether to pursue ("go ahead") or stop the project. If the decision is positive, the aircraft will be offered for sale and the *detailed design phase* will be launched. The goal of the detailed design phase is to design the aircraft for production, hence manufacturing drawings of every single part will finally be available and the manufacturing processes will be implemented. In the end, flight testing is carried out and after successful certification by the authorities, the aircraft enters service (EIS) with the customers.

The above text suggests that the design process is rather rigid and sequential. However, the contrary is true. The process is highly dynamic, iterative and the described phases overlap. Some remarks regarding the time line shall be made. The time between "go ahead" and certification is between four and six years for commercial transport aircraft. Torenbeek (2013) groups the requirement phase, the conceptual design phase and the preliminary design phase into a *configuration phase*. This phase takes two to five years prior to "go ahead". A note from above is valid in this context as well: The process is dynamic and varies depending on the company and the type of aircraft.

At this point the impression could arise that the described process is applicable to a "clean sheet design" only, that is, the product is designed from scratch. This must not necessarily be the case. The described phases are also applied to advance an existing product. Accordingly, the processes and methods of the configuration phase are used to evaluate the impact of technology changes and other modifications on existing aircraft. If they are selected, such modifications are then designed in detail, introduced to production and finally offered to the customer as a derivative of the original product. Recent examples are the Airbus A320neo and Boeing's 737MAX.

The author suggests the following textbooks, on which the above text is based, to be consulted for further information and studies: Gudmundsson (2013),Torenbeek (2013), Howe (2000), Roskam (1989b,a, 1990b,a, 1991, 1997c,b,a).

## 2.2 Aircraft Design Tools at a Glance

The availability of computer systems laid the foundation for software development encompassing the phase of configuration development in aircraft design. Such automated design synthesis systems aim at enabling designers to evaluate a higher number of configurations at a given time. Secondly such systems make the topic of parameter variation and optimization available in configuration development. At the bottom line automated design synthesis programs help to achieve more credible predictions and hence to reduce development risk. The listing in Table B.1 gives an overview of the tools applied in conceptual and preliminary aircraft design, although it does not claim to be exhaustive. For detailed information, the specified literature is to be consulted. Most of the programs originate in academia or are provided by consultancy firms. Information on programs used by aircraft manufacturers is not available to the author. This is reasonable, since such programs contain the expertise and proprietary information of a company. All automated design synthesis systems have in common, that the workflow of the conceptual and preliminary design phases is modelled. Differences are made as to how and up to which level of detail this is done. Programs which mainly use textbook-based methods are e.g. CAPDA, FLOPS, PASS, PIANO, PreSTo and RDS. These tools only require a limited number of input parameters, which is advantageous when exploring a larger design space. Programs which access physics-based methods are, e.g., AAA, ACSYNT, AIDA, CDS, CEASION, MICADO, PrADO and pyACDT. Such systems are commonly used within the preliminary design phase, because of their higher demand of analysis time and input parameters, as compared to the previously mentioned programs.

## 2.3 PrADO - An Aircraft Design and Optimisation Environment

As mentioned above, PrADO (Preliminary Aircraft Design and Optimisation tool) will be used for this work. PrADO is an in-house program of the Institute of Aircraft Design and Lightweight Structures, TU Braunschweig, which covers a wide range of aspects of aircraft preliminary design. The development of PrADO started in the late 1980's and has been continued ever since. In order to work with PrADO, an initial aircraft concept is required. This concept serves as a basis for various modules to calculate all relevant data to assess the design's quality. As a result, an aircraft description, its performance data and properties are obtained. PrADO features three analysis modes: single design analysis, multi parameter variation, and multi-parameter optimization. The basic concept of the program is outlined in Figure 2.2. In order to obtain viable results PrADO has to be configured by the user. The most important decision is to select appropriate analysis methods for each step of the design chain. PrADO therefore contains various modules, marked MDi in Figure 2.2, each of which is designated to fulfil one special task e.g. aerodynamic analysis, estimation of structural mass, mission analysis to estimate fuel masses, etc.

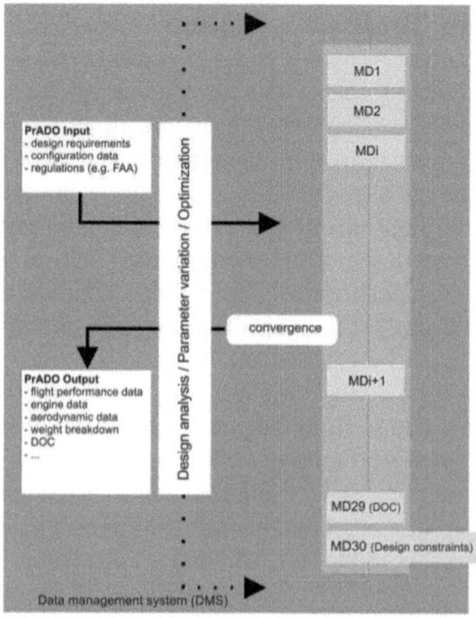


**Figure 2.2:** Basic concept of PrADO

The available methods grouped within those modules range from statistical methods to physics based models. From an aircraft developers point of view PrADO is used within both, the conceptual and the preliminary design phase.
The modular structure of the software allows for the fast integration of additional methods which is beneficial for the work presented here. The following sections contain more detailed expatiations of the aerodynamic module, the structural analysis module and the propulsion module, because the work presented in this thesis focuses on adaptations to them. For further information the author refers to Heinze (1994), Heinze et al. (2001), Österheld et al. (2001) and Hansen et al. (2008).

### 2.3.1 The Aerodynamic Module

The aim of the aerodynamic module is to provide aerodynamic data for the aircraft design process. A basic input to the module is the parametric geometry description provided by the geometry modules of PrADO. Four aerodynamic settings are being analysed. These are a take-off, a cruise, an approach and a landing configuration. The calculated data is fed back into the design process with the help of an aerodynamic performance map and hence available to other modules. The methods currently used for aerodynamic analysis are based on the potential theory. The LIFTING LINE code (cf. Horstmann, 1987), which is based on the lifting-line theory and HISSS (cf. Fornasier, 1985), a panel code, are in use and give results of adequate accuracy for the preliminary design process provided that the flow is mainly attached and subsonic. A higher-fidelity method for aerodynamic data prediction is proposed here which uses the German Aerospace Center's (DLR) CFD code TAU (cf. Schwamborn et al., 2006) to solve the inviscid Euler equations. This has the advantage of being valid for a wide range of Mach numbers (sub-, trans- and supersonic) and geometries and to provide increased accuracy for complex flow problems (e.g. vortical flows). Details on the method were previously published by the author (cf. Schuermann et al., 2014). Viscous drag is estimated by handbook methods.

### 2.3.2 The Structural Analysis Module

Österheld (2003) states that the masses of the load bearing structure accounts for up to 65% of an aircraft's operational empty mass (weight). Consequently, the provision of reliable methods for the estimation of the structural mass of an aircraft is the main intention of the structural analysis module. PrADO provides statistical and semi-empirical methods for this purpose. On the other hand analytical beam-stick models and a finite element process are available. The available structural analysis module, SAM, provides a finite element model and is taken as a starting point in this thesis to develop the SAM2 module. In contrast to SAM, SAM2 implements the parametric geometry description delivered by PrADO more strictly. Furthermore, the discretisation level of the model can be influenced, which requires the use of shell and beam elements accordingly (cf. 4.2). Additionally, the distribution of secondary mass, fuel loading and payload distributions are taken directly from the respective PrADO modules. Loads for ground cases and trimmed flight cases are taken from the PrADO database, whereas the aerodynamic loads are calculated by solving the inviscid Euler equations. The model generation process is implemented into a multi-model generator (MMG). Structural sizing is performed by a newly developed structural sizing module (SSM) which provides various sizing routines and calculates mass and centre of gravity information.

### 2.3.3 The Propulsion Module

The purpose of the propulsion module is threefold. Firstly, the engine is designed on the basis of given input data (on design). Secondly, an engine performance map is calculated (off design) and thirdly the available thrust is compared with the required thrust. The first and the second steps are carried out either by reading from a fixed data set or by thermodynamic cycle analysis. The results are available to other modules at any stage of the design process. PrADO provides various models for the analysis of turbojet, turbofan and turboprop engines. The work presented here uses a newly implemented cycle for a turbofan engine with mixed exhaust flow based on Mattingly et al. (1987) and detailed, in Chapter 3.3.

# 3 Fundamentals

The relevant theoretical aspects presented in the following sections do not claim to be complete but rather provide the necessary background for the upcoming considerations and investigations. The physical principals of aerodynamics, which lead to the Navier-Stokes equations are explained. The inviscid Euler equations are introduced, followed by the explanation of a method to estimate viscous drag. The basic equations of structural mechanics leading to the finite element method (FEM) are illustrated and subsequently aspects of structural sizing are explained. In detail these include failure criteria, sizing strategies and a smoothing algorithm. Furthermore aspects regarding propulsion are outlined. All methods that are introduced in the forthcoming chapters are selected as a compromise between computational efforts and accuracy, where details are explained in the corresponding sections of the text.

## 3.1 Aerodynamic Analysis

The governing equations of fluid dynamics are based on the conservation laws, i.e. the conservation of mass (continuity equation), the conservation of momentum (Newton's second law), the conservation of energy (first thermodynamic law). The resulting five equations are refered to as Navier-Stokes equations[1]. Subsequently, a brief derivation from these equations and the assumptions that lead to the inviscid Euler equations are detailed. The section is being concluded with remarks on aerodynamic drag. The following derivations refer to the textbooks of Anderson (1995), Lecheler (2009), Schlichting & Truckenbrodt (2001) and Wesseling (2005), if not stated otherwise.

### 3.1.1 Navier-Stokes Equations

Consider an infinitesimally small element fixed in space with the fluid moving through it. Applying the principle of conservation of mass to this infinitesimal volume, one obtains equation 3.1. This is the continuity equation in conservation form (also known as divergence form)[2]. In words: The net flow out of the element equals the change of mass inside the element.

$$\nabla \cdot (\rho \mathbf{V}) = -\frac{\partial \rho}{\partial t} \tag{3.1}$$

[1]From a historical perspective solely the three momentum equations are named Navier-Stokes equations. More recent literature refers to the entire set of five equations as Navier-Stokes equations. Throughout this thesis the latter convention is used.

[2]A fine description about the differences between conservation and non-conservation forms of the equations is given by Anderson (1995) and will not be detailed here.

The expression $\nabla \cdot (\rho \mathbf{V})$ denotes the net mass flow out of the element, whereas $\frac{\partial \rho}{\partial t}$ denotes the change of mass inside the element, with $\mathbf{V}$ as the velocity vector, $\rho$ as the fluid's density and $\nabla$ as the vector differential operator.
Newton's second law expresses the second conservation principle (conservation of momentum) as given in equation 3.2.

$$m \cdot \mathbf{a} = \mathbf{F} \tag{3.2}$$

The force vector $\mathbf{F}$ comprises body forces, such as gravity, electric and magnetic forces and surface forces that are viscous forces and pressure forces. The $\mathbf{a}$-vector contains the acceleration to which the mass $m$ is exposed. Keeping this equation in mind and applying the principle of conservation of momentum to the infinitesimal volume, one obtains equations 3.3, 3.4 and 3.5. These are the momentum equations in conservation form, derived for each dimension in a Cartesian coordinate system.

$$\frac{\partial(\rho u)}{\partial t} + \nabla \cdot (\rho u \mathbf{V}) = -\frac{\partial p}{\partial x} + \frac{\partial \tau_{xx}}{\partial x} + \frac{\partial \tau_{yx}}{\partial y} + \frac{\partial \tau_{zx}}{\partial z} + \rho f_x \tag{3.3}$$

$$\frac{\partial(\rho v)}{\partial t} + \nabla \cdot (\rho v \mathbf{V}) = -\frac{\partial p}{\partial y} + \frac{\partial \tau_{xy}}{\partial x} + \frac{\partial \tau_{yy}}{\partial y} + \frac{\partial \tau_{zy}}{\partial z} + \rho f_y \tag{3.4}$$

$$\frac{\partial(\rho w)}{\partial t} + \nabla \cdot (\rho w \mathbf{V}) = -\frac{\partial p}{\partial z} + \frac{\partial \tau_{xz}}{\partial x} + \frac{\partial \tau_{yz}}{\partial y} + \frac{\partial \tau_{zz}}{\partial z} + \rho f_z \tag{3.5}$$

The right term of the equations 3.3 to 3.5 is the total force that is acting on the fluid. It is composed of body forces $f_i$ and surface forces, which are divided into the pressure force $\frac{\partial p}{\partial i}$ and the forces resulting from viscous effects (stresses) $\tau_{ij}$. The indices i and j denote the directions x, y, and z in a Cartesian coordinate system. The left term of the equations 3.3 to 3.5 forms the left part of the equation 3.2. It expresses the acceleration acting on the mass of the fluid and is expressed as the sum of the local derivative and the convective derivative of the fluid's velocity.
The third conservation principle is that of conservation of energy. It states that the rate of change of energy, $\frac{dE}{dt}$, inside the control volume equals the sum of the net flux of heat, $\dot{Q}$, into it and the rate of work, $\dot{W}$, exerted on the element caused by body and surface forces. In short:

$$\frac{dE}{dt} = \dot{W} + \dot{Q} \tag{3.6}$$

Applying the principle of conservation of energy (equation 3.6) to the infinitesimal control volume, one obtains equation 3.7 in conservation form.

$$
\begin{aligned}
\frac{\partial}{\partial t}\left[\rho\left(e+\frac{V^2}{2}\right)\right]+\nabla\cdot\left[\rho\left(e+\frac{V^2}{2}\right)\mathbf{V}\right] &= \rho\dot{q}+\frac{\partial}{\partial x}\left(k\frac{\partial T}{\partial x}\right)\\
&+\frac{\partial}{\partial y}\left(k\frac{\partial T}{\partial y}\right)+\frac{\partial}{\partial z}\left(k\frac{\partial T}{\partial z}\right)-\frac{\partial(up)}{\partial x}-\frac{\partial(vp)}{\partial y}-\frac{\partial(wp)}{\partial w}\\
&+\frac{\partial u\tau_{xx}}{\partial x}+\frac{\partial u\tau_{yx}}{\partial y}+\frac{\partial u\tau_{zx}}{\partial z}+\frac{\partial v\tau_{xy}}{\partial x}+\frac{\partial v\tau_{yy}}{\partial y}\\
&+\frac{\partial v\tau_{zy}}{\partial z}+\frac{\partial w\tau_{xz}}{\partial x}+\frac{\partial w\tau_{yz}}{\partial y}+\frac{\partial w\tau_{zz}}{\partial z}+\rho\mathbf{f}\cdot\mathbf{V}
\end{aligned}
\tag{3.7}
$$

$\rho\dot{q}$ denotes the volumetric heating of the element with $\dot{q}$ as heat flux. The expression $\frac{\partial}{\partial i}\left(k\frac{\partial T}{\partial i}\right)$ denotes the heat flow due to thermal conduction in i-direction with $k$ as thermal conductivity . $\frac{\partial(up)}{\partial x}$ is the rate of work contributed by pressure forces with $u$ as velocity in x-direction and $p$ the pressure. In the y and z directions $u$ is replaced by the velocities $v$ and $w$ respectively. The terms $\frac{\partial\tau_{ij}}{\partial i}$ multiplied by their appropriate velocity denote the rate of work caused by viscous effects, hence shear and normal stresses. As above, i and j denote the directions x, y, and z in a Cartesian coordinate system. The term $\rho\mathbf{f}\cdot\mathbf{V}$ denotes the work done by body forces whereas $\mathbf{V}$ is the velocity vector and $\mathbf{f}$ is the vector of the body forces. On the left side of the equation the change of energy is expressed in terms of total energy $\left(e+\frac{V^2}{2}\right)$, where $e$ denotes the internal energy and $\frac{V^2}{2}$ the kinetic energy. According to Lecheler (2009) potential energy can be neglected for a gaseous medium.

One can observe that the five Navier-Stokes equations contain more unknowns than equations. Consequently, additional relations are needed in order to solve the equations. These relations were developed by Stokes and link the stresses with velocities. For further details on these relations, different forms of the equations and their solution, one might consult the literature on computational fluid dynamics (CFD) (e.g. Anderson (1995); Lecheler (2009); Schlichting & Truckenbrodt (2001); Wesseling (2005)).

### 3.1.2 Reynolds-Averaged Navier-Stokes Equations

The solution of the Navier-Stokes equations is calculated numerically and often after simplifying it further to the Reynolds-averaged Navier-Stokes (RANS) equations. The underlying principle of the RANS equations is Reynolds decomposition, where an instantaneous quantity is split into time-averaged and fluctuating quantities. To close and finally solve the RANS equations the non-linear Reynolds stress term needs modelling. This has led to the creation of turbulence models, which are often given as a set of one or more equations. The result is an approximate time-averaged solution primarily intended to describe turbulent flows. Thorough derivations are given by, e.g., Anderson (1995); Eisfeld (2002); Wesseling (2005). Additionally, one might consult Wilcox (1998) for information on turbulence modelling. The computational cost for the solution of the RANS equations is currently not reasonable within the framework of preliminary aircraft design, hence further simplifications are needed, which lead to the so-called Euler equations described in the following section.

### 3.1.3 Euler Equations

The inviscid Euler equations are computationally less expensive, give adequate accuracy and are therefore more suitable for preliminary design tasks. The term *inviscid* implicitly shows the reason for the reduced analysis time: the boundary layer is not modelled since viscous effects are neglected. In general, this results in an underestimation of drag and an overestimation of lift. The Euler equations are applicable to unsteady, compressible flow problems, hence shocks as they appear in supersonic flow can be captured. As mentioned, the fundamental difference between the (Reynolds-averaged) Navier-Stokes equations and the Euler equations is that terms including viscosity (stresses) and thermal conductivity are neglected. The continuity equation, equation 3.1, does not change. Removing the stresses from the momentum equations (3.3 to 3.5), yields equations 3.8 to 3.10.

$$\frac{\partial(\rho u)}{\partial t} + \nabla \cdot (\rho u \mathbf{V}) = -\frac{\partial p}{\partial x} + \rho f_x \tag{3.8}$$

$$\frac{\partial(\rho v)}{\partial t} + \nabla \cdot (\rho v \mathbf{V}) = -\frac{\partial p}{\partial y} + \rho f_y \tag{3.9}$$

$$\frac{\partial(\rho w)}{\partial t} + \nabla \cdot (\rho w \mathbf{V}) = -\frac{\partial p}{\partial z} + \rho f_z \tag{3.10}$$

The energy equation (3.7) reduces to equation 3.11.

$$\begin{aligned} \frac{\partial}{\partial t}\left[\rho\left(e+\frac{V^2}{2}\right)\right] + \nabla \cdot \left[\rho\left(e+\frac{V^2}{2}\right)\mathbf{V}\right] = \\ \rho\dot{q} - \frac{\partial\,(up)}{\partial x} - \frac{\partial\,(vp)}{\partial y} - \frac{\partial\,(wp)}{\partial w} + \rho\mathbf{f}\cdot\mathbf{V} \end{aligned} \tag{3.11}$$

The five Euler equations have six unknowns expressed as $\rho$, $u$, $v$, $w$, $p$ and $e$. Under the assumption of a perfect gas, the required sixth equation is the equation of state for a perfect gas as expressed in equation 3.12.

$$p = \rho \cdot R \cdot T \tag{3.12}$$

In this equation, $R$ is the specific gas constant and $T$ the temperature. The introduction of $T$ requires an additional relation between the state variables. For a calorically perfect gas, this is equation 3.13 where $c_v$ is the specific heat at constant volume.

$$e = c_v \cdot T \tag{3.13}$$

### 3.1.4 Aerodynamic Drag

Aerodynamic drag is a force acting in the opposite direction of the airspeed vector. Thrust results in a force to overcome drag. Consequently, an aerodynamically efficient design will help to reduce fuel burn and cost, which is one goal amongst others in aircraft design. For this reason, it is necessary to estimate drag at the required accuracy at every stage of the design process. In general, handbook methods, numerical flow simulation and experiments in wind tunnels are used for this task. A detailed overview of drag prediction methods is given by Boppe (1992, p. 7-7 et seq.). Various schemes exist that decompose drag into its components (see, for example, Torenbeek, 2013; Howe, 2000; Roskam, 1990b; Raymer, 2006). Figure 3.1 shows a scheme suggested by Gur et al. (2009), which is used here.

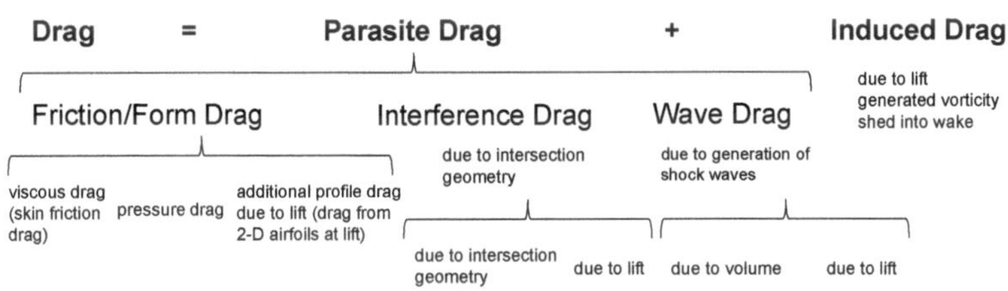


**Figure 3.1:** Decompostion of drag based on Gur et al. (2009)

*Induced drag*, $C_{drag,i}$, is inviscid lift-dependent drag. It is caused by vortices in the wing's wake due to lift. In preliminary aircraft design, induced drag is often derived by assuming a quadratic polar ($C_{drag,i} \sim C_{lift}^2$) or from methods based on potential theory (cf. Horstmann, 1987; Erickson, 1990; Fornasier, 1985). *Parasite drag* is subdivided into *interference drag*, *wave drag* and *friction/form drag*. It is not equal to zero drag since lift-dependent drag components are included, which is not the case in the definition of zero drag. Interference drag originates from the intersection of the vehicle's components (cf. Tétrault, 2000). Wave drag is caused by shock waves that appear at speeds higher than the critical Mach number. Wave drag arises in the transonic regime and continues to exist in supersonic flight. Boppe (1992) states that wave drag *"develops as a result of differences in the compressibility of air in subsonic and supersonic flow"*. Friction and form drag can be further divided into *pressure drag /form drag*, *additional profile drag due to lift* and *skin friction drag*. Pressure drag is a function of a body's geometry. It is caused by eddying motions set up in the fluid by the passage of the body. Mathematically spoken, the normal pressure integration parallel to the freestream flow is non-zero. Profile drag due to lift is caused by the change of the boundary layer thickness. Skin friction drag is caused by shear stresses in the boundary layer. It is not computed by solving the Euler equations. However, the contribution of skin friction drag/viscous drag to overall aircraft drag is not negligible (Howe, 2000, pp. 19,145).

A common drag estimation approach in preliminary aircraft design are *Component Buildup Methods*. Their principle is to divide an aircraft into components and to estimate drag for each component with appropriate formulae before adding it all up to total drag. Interference drag is accounted for by factors derived from statistics and experiments. In general, friction drag is estimated based on a flat plate analogy. Pressure drag is accounted for by form factors. Similar approaches are used for other types of drag. For a detailed description on drag estimation one should consult e.g. Hoerner (1965) or LTH (2006). For details on the method to estimate viscous drag used here, see Appendix C.1. For theoretical derivations of drag and its physical principles the reader is referred to Boppe (1992); Hoerner (1965); Schlichting & Truckenbrodt (2001) and White (2005).

## 3.2 Estimation of Structural Mass and Structural Sizing

The primary goal of structural analysis within the framework of preliminary aircraft design is to estimate the structural mass and the centre of gravity of a design exposed to a given set of loads and restrictions. The basis of the analysis is an appropriate analysis model. Statistical analysis models provide fast means to derive mass properties for a design. They work well for designs that are similar to those covered by the statistical data, however they will give poor results for designs that are far away from what is in the statistical data set. Heinze (2007) labels such methods as *class one methods*. Sticking to that notation, simple analysis models, e.g. beam-stick models, serve as a basis for *class two methods*. These models have limited sensitivity to parameter changes in the design but are computationally not expensive. *Class two methods* are most frequently physics-based and are often calibrated by statistical data restricting their use to a certain extent to configurations covered by the statistical data set. Finally, a third set of methods is defined, i.e. *class three methods*. The underlying analysis model is always physics-based, e.g. a finite element model, and a more detailed structural layout can be considered. These models require higher computational and modelling effort but are sensitive to parameter changes in the design space. Both, *class two methods* and *class three methods* require structural sizing. The above reasons have influenced the decision to use a finite element model, hence a *class three method*, for the work presented here. Finite element models are meanwhile well established in aircraft pre-design and increased computational performance allows their extensive use. Often, a realistic representation of the deformation of the structure is desired. This can be achieved by the use of a finite element model and automated modelling has become a reliable asset in preliminary aircraft design. For details on the modelling see Chapter 4.
The scope of the structural analysis model is to show the response of the structure to applied loads. The result is being fed into a sizing model which adapts the design variables (e.g. stiffness, skin thickness) to satisfy selected criteria and to comply with given restrictions. The primary goal typically consists in minimizing mass.

Restrictions are mechanical strength, stiffness and deformation, whereas this listing does not claim to be complete. Sizing is ensured either by means of an optimization strategy (cf. Schumacher, 2005), or by applying optimality criteria methods, such as the method of fully stressed design (FSD)[3]. The former generally requires a much higher computational effort than the latter, which is the reason why the latter is frequently used in preliminary aircraft design and throughout the work presented here. Amongst others, Harzheim (2008, chap. 5) states that optimality criteria methods lack general validity, hence they are in most cases tailored to be applicable to a specific class of problems only. Haftka & Gürdal (1992, chap. 9) define the following two steps for optimality criteria methods:

1. The definition of optimality criteria and
2. The resizing of the structure in order to satisfy the optimality criteria

The subsequent sections will introduce the fundamentals of the finite element approach in brief (Section 3.2.1), followed by the introduction of fully stressed design in combination with the v. Mises yield criterion as optimality criterion for ductile isotropic materials in Section 3.2.2. Failure criteria for brittle materials as optimality criteria for composite material are outlined in Section 3.2.3. A resizing algorithm developed within the scope of this work will be explained in Section 3.2.4 followed by the introduction of a smoothing algorithm in Section 3.2.5.

### 3.2.1 The Finite Element Method

This section is to provide a short introduction into the finite element method. For thorough derivations and examples one might consult literature as for example Zienkiewicz & Taylor (1989) and Bathe (2002).
Mechanical Systems with an infinite number of degrees of freedom are referred to as continuous. The behaviour of such a continuum is described by partial differential equations for which only a limited number of analytical closed-form solutions exist. Common practice is therefore to use numerical methods, such as the finite element, finite boundary or finite difference method to solve the problem. The finite element method is a state-of-the-art approach in engineering. The continuous problem is transformed into that of a finite number of discrete problems in order to obtain an approximate solution. Nodes and elements ensure the discretisation, whereas desired quantities are calculated at the former, and the latter provide interpolation schemes for such quantities in other areas of the discretised continuum. The element formulation is problem-dependent and based on physics. The basic assumption of the finite element approach is that a constrained structural system fixed at some point in space, will deform when subjected to forces. The principle of minimum total potential energy states that such a system deforms or moves to a position that

[3]The method of fully stressed design is related to the stress-ratio method

minimizes the total potential energy. Losses are dissipated as heat. In the structural mechanical setting as used here, the quantities calculated at the nodes are displacements. Applying this principle results in equation 3.14 as the basic equation of a discretised finite system when assuming linear theory (small deformations). In this equation $\mathbf{M}$ is the mass matrix, $\mathbf{K}$ denotes the stiffness matrix and $\mathbf{f}$ is the load vector. The $\mathbf{u}$-vector contains all translational and rotational displacements, whereas $\ddot{\mathbf{u}}$ contains the derivatives of $\mathbf{u}$ with respect to time, hence the accelerations. The equation is solved for $\mathbf{u}$. The displacements are further used to calculate, e.g. stresses and strains.

$$\mathbf{M}\ddot{\mathbf{u}} + \mathbf{K}\mathbf{u} = \mathbf{f} \tag{3.14}$$

For static analysis $\ddot{\mathbf{u}}$ becomes the null vector and equation 3.14 can further be reduced to equation 3.15.

$$\mathbf{K}\mathbf{u} = \mathbf{f} \tag{3.15}$$

In the context of this work, the commercial finite element program MSC.NASTRAN® is used to solve the above equations. The reader is advised to consult the *NASTRAN® Linear Static Analysis User's Guide* (Anon, 2011) for more information.

### 3.2.2 Fully Stressed Design

The underlying thought of the concept of fully stressed design is that the load of a member of the structural model is constant during the resizing of the structure (cf. Adelman et al., 1980). Additionally, it is implied that adding and removing material from a structural member that is not fully stressed will primarily influence the stress of that member. The notion of a *fully stressed* structural member finds its mathematical expression in equation 3.16. In this context, fully stressed means that the value of $FI$, the so-called failure index[4] is exactly one for at least one loading condition. A value above one indicates failure of the structural member, whereas a value below one indicates a structural member on which more material has been invested than theoretically necessary. Hence, it is desired to achieve a value of or close to one.

$$FI = \frac{|\sigma_{load}|}{\sigma_{allow}} \qquad \begin{cases} < 1.0 \\ \equiv 1.0 \\ > 1.0 \end{cases} \tag{3.16}$$

The term *theoretically* requires further explanation, but prior to that it is to be noted that $\sigma_{load}$ and $\sigma_{allow}$ in equation 3.16 denote the stress to which an element is

[4]The concept of failure indices is closely linked to composite failure criteria. However, it is introduced to equation 3.16 in order to provide a consistent terminology within this text.

exposed due to loading and the allowable stress based on material data and knock-down factors $\phi_M$ as calculated in equation 3.17.

$$\sigma_{allow} = \phi_M \cdot strengthvalue \tag{3.17}$$

It might be desirable that in cases of violation of the above implications, structural members are not fully stressed but still fulfil the requirement of their corresponding $FI$ to be below one. These structural members help alleviate stress from other members and might consequently help to find a better design. At this stage, the reason for the introduction of the term *theoretically* earlier in the text becomes obvious. However, is the result the optimum? The answer to that question is multifaceted. Firstly, for all statically determinate systems the assumption that adding material or removing material from a structural member will have an effect on the stress of that member is correct. It has been proven that the result is the optimum design with respect to weight (cf. Michell, 1904). Secondly, for statically indeterminate systems and systems with more than one load case the optimum with respect to weight is not necessarily fully stressed (cf. Haftka & Gürdal, 1992; Harzheim, 2008), however in most cases the result is close to the optimum and the procedure becomes one of iterative nature.

For rod elements, $\sigma_{load}$ is the axial stress and equation 3.16 stays valid. Beam elements are sized taking axial stress and maximum bending stress into account. Two-dimensional structural members are usually considered to be under plain stress. This is a valid assumption for the work presented in this text, since aircraft structures are thin in comparison with their areal dimensions. For these members composed from isotropic material and under multi-axial loading, $\sigma_{load}$ is frequently based on the v. Mises yield criterion for plain stress and derived as in equation 3.18.

$$\sigma_{load} = \sigma_v = \sqrt{\sigma_x^2 + \sigma_y^2 - \sigma_x\sigma_y + 3\tau_{xy}^2} \tag{3.18}$$

Here $\sigma_x$ and $\sigma_y$ denote the stresses in the x- and y-directions of the element coordinate system, whereas $\tau_{xy}$ is the shear stress in the xy-plane of the element coordinate system. The optimality criterion of fully stressed design is not suitable for composite materials. It converges slowly *"because of the redundancy inherent in the use of composite materials"* (Haftka & Gürdal, 1992, p. 352). Therefore, the next section will introduce criteria that are applicable to composite materials.

### 3.2.3 Failure Criteria for Composite Material

The right use of failure criteria makes the method applicable to composite materials. Common composite materials are of orthotropic nature, that is, they have a minimum of two orthogonal axes of rotational symmetry. In general, their mechanical properties are different along each axis: hence, such materials are anisotropic. Basically, two types of criteria can be distinguished. On the one hand, there are global fracture criteria. According to Schürmann (2008), these criteria are formulated by a single polynomial and predict failure without making a statement about its nature (i.e. matrix failure or fibre failure). On the other hand, action-plane criteria like the one proposed by Puck (1996) make a distinction between the failure modes. An overview and evaluation of failure criteria is given within the World Wide Failure Exercise (WWFE). The results are summarized by Soden et al. (2004). Based on the first edition of the WWFE, a second part has been completed (see Kaddour & Hinton, 2013), and a third part is in progress (cf. Kaddour et al., 2013). A critical discussion regarding the methodology of the WWFE by Christensen (2013) is also worthy of mention.

As outlined before failure is characterized by failure modes, of which the two basic types are fibre failure and matrix failure. For many engineering applications, such as aircraft structures, it is assumed that laminate failure is dominated by fibre failure and that laminate failure occurs as soon as the fibres in one ply fail. This is known as first ply fibre failure (FPFF) and a conservative approach. FPFF assumes that the fibres are able to carry the load even if matrix failure has already occurred[5]. The assumption of first ply fibre failure is valid in the context of preliminary aircraft design since knowledge about the global characteristics of the aircraft is desired and no detailed design on a component level is made. Therefore and according to Rieke (2013), information about failure modes is not useful in the framework of aircraft pre-design.

The large number of criteria and the previously made comments raise the question as of which criteria are suitable for preliminary aircraft design. In this text, a maximum strain criterion (Hart-Smith, 1998) and an approach by Liu & Tsai (1998) are presented. For a summary of the criterion proposed by Puck (1996), the well-disposed reader is advised to consult Rieke (2013). This source shows that the criteria by Puck and by Liu & Tsai (1998) are most suitable for aircraft pre-design, even though the additional information about fracture modes given by the former is not used in the preliminary aircraft design environment.

Equation 3.19 shows a maximum strain criterion. Strains of a ply are compared against allowable values in this simple criterion.

$$FI = \max\left\{\frac{|\epsilon_1|}{X_{t,c,(s)}}, \frac{|\epsilon_2|}{Y_{t,c,(s)}}, \frac{|\gamma_{12}|}{S_s}\right\} \tag{3.19}$$

The values of $\epsilon_{1,2}$ and $\gamma_{12}$ denote strains in a ply in 1-direction, 2-direction and for shear strain in the 1-2 plane. The allowable strain values are marked $X_{t,c,(s)}$ for longitudinal tensile and compressive strengths, $Y_{t,c,(s)}$ for transverse tensile and compressive strengths, and $S_s$ for shear strength in the 1-2 plane.

The quadratic failure criterion by Tsai, as proposed in Liu & Tsai (1998) is based on stresses and does, unlike the previous strain criterion, take interaction between stresses into account.

[5]For ultimate strength matrix failure is acceptable. However, matrix failure leads to a decrease of strength.

$$FI = \frac{\sigma_1^2}{X_t \cdot X_c} + \frac{2 \cdot F_{xy}^* \sigma_1 \sigma_2}{\sqrt{X_t X_c Y_t Y_c}} + \frac{\sigma_2^2}{Y_t \cdot Y_c} + \frac{\tau_{12}^2}{S^2} + \left[\frac{1}{X_t} - \frac{1}{X_c}\right] \sigma_1 + \left[\frac{1}{Y_t} - \frac{1}{Y_c}\right] \sigma_2 \quad (3.20)$$

The values of $\sigma_{1,2}$ and $\sigma_{12}$ denote stresses in a ply in 1-direction, 2-direction and for shear strain in the 1-2 plane. The allowable stress values are marked $X_{t,c}$ for longitudinal tensile and compressive strengths, $Y_{t,c}$ for transverse tensile and compressive strengths, and $S$ for shear strength in the 1-2 plane. $F_{xy}^*$ denotes the normalized interaction term of a quadratic failure criterion that is to be determined by experiments. Should experimental data not be available, a value of $F_{xy}^* = -0.5$ is suggested.

### 3.2.4 The Resizing Algorithm

As outlined in the introduction to Section 3.2, resizing algorithms complement optimality criteria methods. Their purpose is to change the value of design variables in order to achieve a feasible design under a set of restrictions. Besides the allowable stresses and strains that are used in the optimality criteria, minimum gage constraints, $a_{min}$, are applied according to equation 3.21, where $a$ denotes the design variable.

$$a_{new} = \max\left(a_{new}, a_{min}\right) \quad (3.21)$$

On the one hand, these constraints serve to assure a technically feasible design, on the other hand, they provide means to account for effects that have been neglected as, for example, damage tolerance or stability effects, should they not have been accounted for in the stress or strain allowables or analysed elsewhere. Optional maximum gage constraints are imposed additionally. As with the smoothing algorithm, (cf. Section 3.2.5), these constraints help to avoid accumulation of mass in regions where it is not meaningful. The following sections provide information on resizing algorithms for various types of structural members.

#### Resizing of Isotropic Shells and Shear Panels

The resizing algorithm for isotropic shells and shear panels is expressed as linear relation between the old and the new thickness of the respective member. In equation 3.22, the thickness of the n-th sizing step, $t_n$, is related to the thickness of the (n-1)-th sizing step, $t_{n-1}$, by multiplication with a failure index $FI$ and a calibration factor $CF$. This equation expresses the basic idea of fully stressed design, or more specifically the stress-ratio method, as explained in Section 3.2.2. The calibration factor helps drive the design towards convergence. The value of $CF = 0.98$ has proved to be appropriate (cf. Section 5.2) for the application in this thesis.

$$t_n = FI \cdot CF \cdot t_{n-1} \tag{3.22}$$

Österheld (2003) suggests a modified sizing algorithm as depicted in equation 3.23. Here, sizing is subject to a damping function that is composed of the square root function and damping coefficients $c_1$ and $c_2$ respectively. Österheld proposes a value of 0.99 for $c_1$ and 1.1 for $c_2$. High values of $c_1$ and $c_2$ result in a rapid change of the thickness, which influences convergence significantly. It is to be noted that the formula proposed by Österheld has been developed for two-dimensional elements that have membrane stiffness only, whereas the elements used for this work have membrane and bending stiffness.

$$t_n = t_{n-1} \cdot \begin{cases} CF \cdot c_1 \cdot \sqrt{FI} & \text{for } FI < 1.0 \\ CF \cdot \left(c_2 + \sqrt{(FI)^2 - 1}\right) & \text{for } FI > 1.0 \\ 1.0 & \text{for } FI \equiv 1.0 \end{cases} \tag{3.23}$$

In contrast to the methodology proposed by Österheld (2003), the calibration factor $CF$ is introduced in this thesis.

The behaviour of the sizing functions is shown in Figure 3.2. It can be seen that the linear method does not differentiate between adding and removing material, whereas the method proposed by Österheld does. The coefficients $CF$, $c_1$ and $c_2$ influence the functions with respect to their position on the ordinate.

### Resizing of Orthotropic Shells

Multiple resizing schemes for orthotropic shells are to be considered. A first scheme considers the thickness of each layer as calculated according to equation 3.22. The resulting thickness of the layer is taken “as is”. This could change a laminate's behaviour drastically, since symmetry conditions might not be respected. On the other, hand each layer is utilized as required.

Rieke (2013) proposes a different scheme. The suggestion is to scale the thickness of each layer by the thickness ratio of the layer with the maximum failure index according to equation 3.24. Here, $t_n(FI_{max})$ denotes the thickness of the layer with the maximum failure index of the n-th sizing step, whereas $t_{n-1}(FI_{max})$ is the thickness of the layer with the maximum failure index of the (n-1)-th sizing step.

$$ratio = \frac{t_n(FI_{max})}{t_{n-1}(FI_{max})} \tag{3.24}$$

The new thickness of the laminate is determined by applying equation 3.25 to each i-th layer of the laminate. Here again, the calibration factor $CF$ is introduced, in contrast to the original method. This modified method is used within this work.

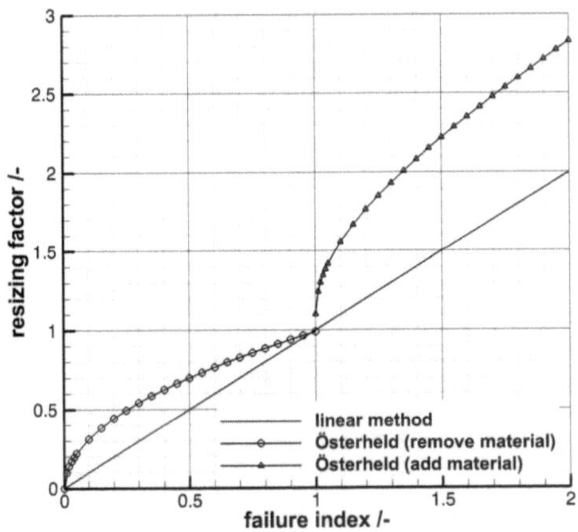


**Figure 3.2:** Influence of the damping function for shell sizing

$$t_{i,n} = ratio \cdot CF \cdot t_{i,n-1} \tag{3.25}$$

This method assures that the characteristics of the laminate (e.g. symmetry) are respected. Furthermore, it is guaranteed that the layer with the maximum failure index is sized exactly as needed.
Schuermann et al. (2013) propose an alternative method in which the thickness ratio of each layer with respect to the total thickness of the laminate is used to size the layer. As with the method proposed by Rieke, the characteristics of the laminate (e.g. symmetry) are respected. However, it is not guaranteed that the layer with the maximum failure index is sized exactly as needed, which may lead to a number of more sizing iterations.

### Resizing of Rods

Rod elements are one-dimensional elements of circular cross-section that carry axial load. Some formulations allow for torsion [6]. The rod elements used within this thesis are subject to axial load only. The applied resizing scheme for rod elements is based on equation 3.22. The calibration factor $CF$ is set to one and the thickness

[6] A more general expression is *bar*. The loading of a bar is the same as for a rod but a bar can be of arbitrary cross section (Donaldson, 1993, p. 231). This definition will be valid within this text.

is replaced by the cross-sectional area $A_n$ and $A_{n-1}$. The result is equation 3.26. This equation would lead to a fully stressed member if a system were not redundant and only subject to one load case (cf. Section 3.2.2). This resizing algorithm is frequently presented in literature for the sizing of truss structures.

$$A_n = FI \cdot \cdot A_{n-1} \tag{3.26}$$

### Resizing of Beams

The fundamental distinction between beams and rods/bars is that the former allow for the application of bending moments and some formulations for shear, while the latter do not (Donaldson, 1993, p. 231). Beams are of arbitrary cross-section. Accordingly, the resizing algorithm for beams takes both, axial and bending stress into account. In sections where axial stress is predominant, the beam's cross-sectional area is sized. For sections with high bending stresses, the area moment of inertia is sized where the most critical value is taken into account. A stress ratio $r_b$, is introduced in equation 3.27 to differentiate between the two cases. Here, $\sigma_b$ denotes bending stress and $\sigma_{ax}$ denotes axial stress.

$$r_b = \frac{\sigma_b}{\sigma_{ax}} \tag{3.27}$$

Sizing of a member is carried out as shown in equation 3.28, where $I_n$ denotes the area moment of inertia of the n-th sizing step, $I_{n-1}$ is the area moment of inertia of the (n-1)-th sizing step and the cross-sectional areas are $A_n$ and $A_{n-1}$ respectively. The *limit* for $r_b$ is imposed based on engineering judgement. In this context a *limit* of 0.5 seems to be appropriate. The calibration factor, $CF$, is set to a value of 1.0.

$$\left.\begin{array}{c} I_n \\ \\ A_n \end{array}\right\} = FI \cdot CF \cdot \left\{\begin{array}{ll} I_{n-1} & \text{for } r_b > limit \\ \\ A_{n-1} & \text{for } r_b \leq limit \end{array}\right. \tag{3.28}$$

## 3.2.5 Smoothing of Design Variables

A smoothing algorithm is integrated in addition to the resizing algorithm. This is necessary for two reasons. First the resizing algorithm tends to create chequerboard patterns for 2D Elements if each element is sized separately. Secondly material is accumulated in regions where it is not meaningful from a design perspective. The reason for both phenomena is that adding material to an element will increase its stiffness and in return "attract" higher stresses and hence more material. The algorithm is a simple chequerboard suppression algorithm for evolutionary structural optimization as proposed by Li et al. (2001) and used by Österheld (2003) for such problems. Equations 3.29 and 3.30 show the relation

$$\alpha_e = \frac{\sum_{i=1}^{n_e} v_i \cdot w_i}{\sum_{i=1}^{n} w_i} \tag{3.29}$$

$$\overline{v_1} = v_1 \cdot \beta_g + \alpha_e \cdot (1 - \beta_g) \tag{3.30}$$

where $n_e$ denotes the number of elements, $\alpha_e$ is the averaged influence coefficient, $v_i$ denotes a value of the i-th connection (neighbouring) element and $w_i$ its influence coefficient. $\beta_g$ is a global influence coefficient, $v_1$ the value which is to be smoothed and $\overline{v_1}$ the smoothed result. In the context of the sizing procedure presented herein, the element's thickness is smoothed ($v_i$) using its surface area as influence coefficient ($w_i$). The global influence coefficient is set to a value between 0.4 and 0.6 (cf. Österheld, 2003).

## 3.3 Propulsion

The purpose of the engine is to deliver sufficient thrust to propel the aircraft. Furthermore different forms of energy, such as bleed air or electrical power, are provided. The notion *power plant*, *propulsion unit* and *engine* are often used synonymously. Within the framework of preliminary aircraft design, information about the performance characteristics of the engine is desired. Parameters of interest are thrust, specific fuel consumption ($SFC$) and emission data under multiple operating conditions. Additionally, the mass of the propulsion system is of interest, and its influence on aerodynamic characteristics is based on its integration into the aircraft.
Two methodologies can be followed to provide the required data. *"Either an engine fixed in size and thrust, fixed-engine, or a scalable rubber engine, can be used for this purpose"* (Schuermann et al., 2015). The use of the former is reasonable if the type and size of the engine is known. This is the case when an existing aircraft is evaluated or new technology in an existing design is rated. For new designs the selection of an engine of the shelf reduces the development costs but is often paid for by performance characteristics that are not precisely matching. The methodology of *rubber engine sizing* provides flexibility in that the engine's thrust and size are matched to the aircraft's requirements. The result is the basis either for the selection of an existing engine, which reduces development costs, or for the development of a new engine, which entails higher costs. Rubber engine sizing requires modelling of the thermodynamic cycle whereas the fixed engine methodology uses such a model or data delivered by the engine manufacturer. Both methodologies are available in PrADO.
Should the model of a thermodynamic cycle be available, it has to be run in two modes. Firstly, the on-design mode is executed. The engine is designed in this mode based on performance parameters, design limitations and flight conditions, which lead to specific design choices and consequently to an engine layout. Input values are parameters such as pressure ratios of components and the conceptual

layout of the engine, e.g. the number of spools. The derived engine layout is frozen and available for further analysis. Finally, the thermodynamic data is fed into the mass estimation process of the propulsion unit.
Secondly, the off-design mode is used to calculate engine performance data for any desired combination of input parameters, giving an understanding on how the engine performs under operating conditions. This process is iterative in many ways. *"The results, such as specific fuel consumption, thrust and component masses are fed back into the overall design loop"* (Schuermann et al., 2015). A remark on supersonic operations: All conventional ducted (airbreathing) jet engines[7] do have in common that the velocity of the flow that reaches the fan has to be below the speed of sound. For supersonic operations, this requirement is normally fulfilled by a specifically designed engine inlet (Diffuser). The determination of specific fuel consumption and other parameters does not differ from engines operated in subsonic mode.
The following section provides an introduction into the basics of a mixed-stream turbofan engine. The model allows for both, fixed and rubber engine procedures. For details on engine concepts and specifics of the thermodynamic cycle, the well-disposed reader is advised to consult the textbooks of Bräunling (2004), Flack (2005) and Mattingly et al. (1987) to which the following information refers, if not stated otherwise.

### 3.3.1 Mixed-Stream Turbofan Engine

Part of the text in the subsequent sections has been published previously by the author in Schuermann et al. (2014). The engine used herein is a turbofan engine with mixed exhaust flow. A schematic representation is shown in Figure 3.3. The letters S and R denote stator and rotor respectively. The core (gas generator) of this engine consists of a high-pressure (HP) compressor, the combustion chamber (combustor), a high-pressure turbine and a low-pressure (LP) turbine. A fan is mounted on the low-pressure spool, driven by the low-pressure turbine. In addition to the fan, a low-pressure compressor can be mounted, which is the case in Figure 3.3. For the analysis conducted here, the fan functions as a de facto low-pressure compressor and no additional low-pressure compressor is mounted. The high-pressure turbine drives the high-pressure spool and hence the high-pressure compressor. A mixer is integrated right after the low pressure turbine, making sure that the bypass stream that passes through the bypass duct and the core stream are mixed before exiting through the nozzle. The nozzle is a convergent-divergent nozzle (de Laval nozzle) with an adaptable cross section. The mixer and the nozzle, which are specific to this engine type, will be explained here in greater detail.

[7] The term *conventional* in this context relates to engines with a compressor, a combustion chamber and a turbine

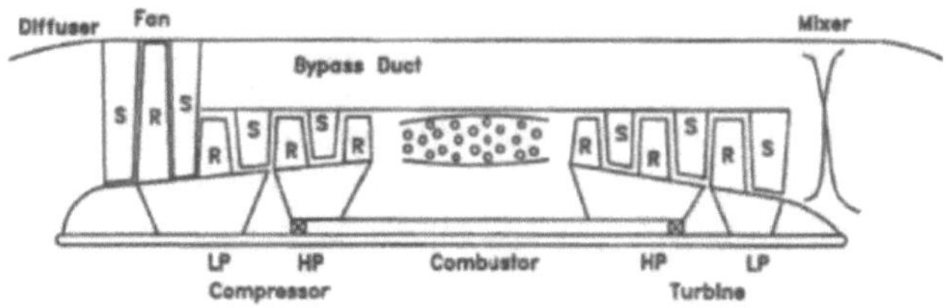


**Figure 3.3:** Schematic representation of a mixed-stream turbofan engine (Flack (2005), p.24)

### The Mixer

The mixer is a constant area mixer. Its purpose is to mix the core gas flow stream with the bypass flow stream before it exits through the nozzle. For the design of the mixer it is required that the Kutta condition is satisfied at the entry of the mixer, hence that the pressure of the core gas flow stream ($p_5$) and the bypass flow stream ($p_5'$) are equal. This implies, in contrast to separated exhaust flows, that the flows can no longer be analysed separately. The situation and station numbers are shown in Figure 3.4. In addition to the above constraint, the Mach number of the flow is restricted. Neither the Mach number of the core gas flow stream ($Ma_5$) nor the Mach number of the bypass flow stream ($Ma_5'$) are allowed to be greater than one (choked) and less than zero (reverse flow) at the entry of the mixer. Mattingly et al. (1987) state that low values for $Ma_5$ and $Ma_5'$ will result in an increase in the engine's cross-sectional area, which is logical having the continuity equation in mind.

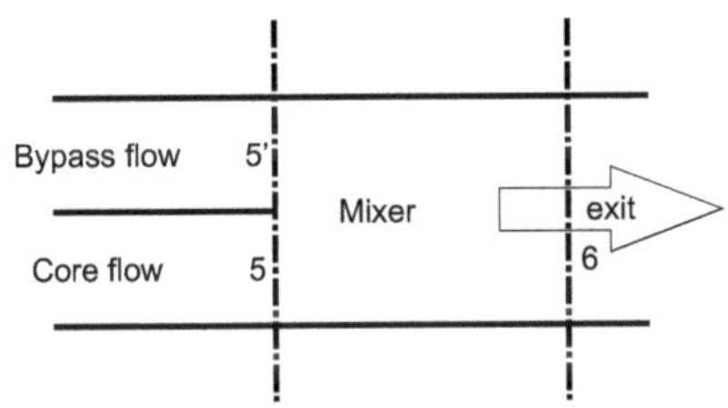


**Figure 3.4:** Station numbers of the mixer

The ideal pressure ratio (no wall friction, no mixing losses) of the mixer reduces to

$$\pi_{Mideal} = f(\alpha', \tau_M, \frac{A_5'}{A_5}, Ma_5, Ma_6, \gamma_5, \gamma_6, R_5, R_6) \tag{3.31}$$

In equation 3.31 $\alpha' = \frac{m_5'}{m_5}$ denotes the bypass ratio of the mixer and $\tau_M = \frac{T_6}{T_5}$ is the temperature ratio of the mixer. $Ma_5$ and $Ma_6$ are the Mach numbers at stations 5

and 6 respectively, whereas $\gamma_5$ and $\gamma_6$ denote the ratio of specific heats and $R_5$ and $R_6$ are the gas constants, all at stations 5 and 6. $\frac{A_5'}{A_5}$ denotes the cross-sectional ratio of the core gas flow stream to the bypass flow stream.
Within the normal on-design loop, only $Ma_6$ needs to be determined because all other values are given from the analysis of previous components. According to Mattingly et al., wall friction can be added by multiplying $\pi_{Mideal}$ with $\pi_{Max}$, which is *"the mixture total pressure ratio due to wall friction only"* (Mattingly et al., 1987, p. 113).

### The Nozzle

The main purpose of the nozzle is to relax the flow to ambient pressure and convert its energy into velocity, hence thrust. For subsonic flow this is carried out by means of a convergent nozzle. High exhaust velocities (with afterburner use often higher Mach one) require a special design of the nozzle. A so-called convergent-divergent nozzle (de Laval nozzle) is used for this purpose. Figure 3.5 shows a sketch of a Laval nozzle and the qualitative behaviour of pressure, temperature and velocity along the nozzle. Sonic flow is reached at the narrowest section of the nozzle ($A_8$), called throat, and then further expanded to velocities higher than Mach one until the nozzle's exit ($A_9$). The augmentation of velocity is accompanied by a drop in temperature and pressure. Since operating conditions are not always design conditions, a nozzle with variable geometry is used to adapt its shape for optimum thrust depending on the working state of the engine (Flack, 2005). The nozzle proposed for use within the scope of this document can be adapted at two positions:

- For supersonic conditions, the cross-section of the throat ($A_8$) is always adapted for the velocity to reach Mach one, while the adjustment of the cross-section at the exit of the nozzle ($A_9$) aims to fulfil the above-mentioned pressure condition, which is that ambient pressure will be reached at $A_9$.

- For subsonic flow, both cross sections are varied in a way that Mach one is reached at the exit of the nozzle ($A_9$) and the pressure condition is satisfied.

The use of a variable nozzle and its advantage of a more flexible adaptation to operating conditions is normally penalized by an increase in weight, when compared with a standard nozzle arrangement. This is the reason why this type of nozzle is hardly found in civil use. For the supersonic business jet, a variable nozzle is indispensable due to the combination of subsonic and supersonic operating conditions under which the engine will efficiently operate.

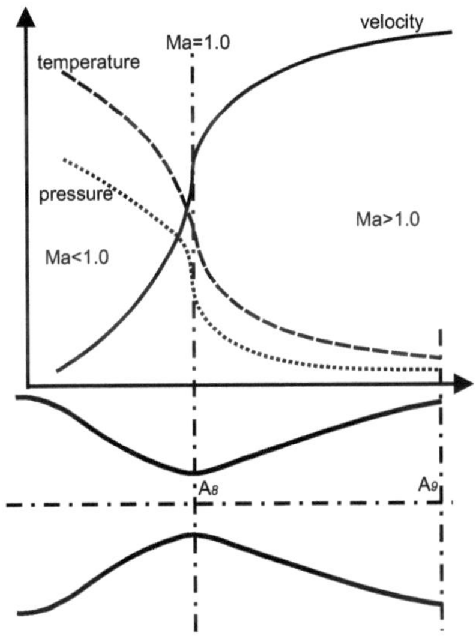


**Figure 3.5:** Convergent-divergent nozzle (bottom) and qualitative state variables (top)

# 4 Modelling Aspects

As outlined above, the aircraft design process is multidisciplinary, thus requiring analysis models for various engineering disciplines. Based on Chapter 3 the subsequent sections aim at giving insight into relevant details regarding such analysis models. Details on the aerodynamic model are given in Section 4.1 followed by information on the structural model in Section 4.2. Section 4.3 provides information on load cases and their derivation.

## 4.1 Aerodynamic Model

The methods currently used for aerodynamic analysis within PrADO are LIFTING LINE (cf. Horstmann, 1987) and the panel code HISSS (cf. Fornasier, 1985). These methods provide results of adequate accuracy for the preliminary design process if the flow is mainly attached and subsonic.
HISSS is equally capable of computing supersonic flow. It works well if the body under consideration is fully immersed within a region of influence. This region of influence is determined by the free-stream Mach cone under the assumption of a linearised supersonic flow formulation. If a part of the body lies outside the zone of influence, the theory, and hence the code, would cease to be applicable[1]. This can happen for two reasons, the first of which is that the free-stream Mach number increases and, as a result, the free-stream Mach cone angle decreases. Consequently, some areas of interest would lie outside the Mach cone, causing the theory to fail. The relation of the free-stream Mach number, $Ma_\infty$, and the Mach cone angle, $\mu$, is given in equation 4.1.

$$\begin{aligned} \sin\mu &= \tfrac{1}{Ma_\infty} \\ \tan\mu &= \tfrac{1}{\sqrt{Ma_\infty^2-1}} \end{aligned} \tag{4.1}$$

The second reason is of geometrical nature. When the body is relatively thick and a part of the boday lies outside the zone of influence, most supersonic panel methods would fail, so does HISSS. Figure 4.1 illustrates the scenario in a simplified manner. In subplot 4.1(a), the leading edge of a wing lies inside the Mach cone, whereas subplot 4.1(b) shows the leading edge outside the Mach cone. The leading-edge sweep angle is denoted $\varphi$, while $\gamma$ is its complement to ninety degrees and $\mu$ the already defined Mach cone angle.
A common expression for panels exposed outside the zone of influence is that of *super-inclined boxes* or *super-inclined panels*. In both cases, failure is caused by

[1]This limitation to the linear potential flow theory is also applicable to other panel codes.

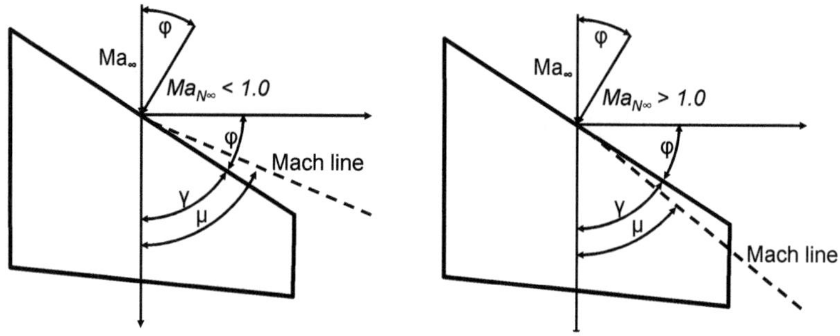


(a) Subsonic leading edge (b) Supersonic leading edge

**Figure 4.1:** Leading edge characteristics

numerical singularities in the supersonic aerodynamic influence coefficient computation. Its details are not of interest for the following text, and literature is to be consulted for more information.
For the above reasons and following suggestions by Mason et al. (1998), Tidd et al. (1992) and others, a higher fidelity method for aerodynamic data prediction is proposed by the author (cf. Schuermann et al., 2013, 2014, 2015) and applied here. This model uses the German Aerospace Center's (DLR) CFD code TAU to solve the inviscid Euler equations. This has the advantage of

a) being valid for a wide range of Mach numbers (sub-, trans- and supersonic) and geometries and

b) provide increased accuracy for complex flow problems (e.g. vortical flows)

Mesh generation and, compared to the currently used methods, high computation time are two challenges to tackle. The first is achieved in an automated process using the CPACS file format (Nagel et al., 2012), which is an XML-based aircraft description format. PrADO and the TAU aerodynamic module exchange relevant aircraft configuration data using this format. *"The geometry is read in by the SUMO grid generator (Kaenel et al., 2008), which automatically creates a tetrahedral mesh"* (Schuermann et al., 2015). Additionally a smoothing operation on the grid is performed with *TAU utilities* in order to increase the mean quality of the mesh. Analysis parameters are configured in order to reduce the computation time of the process to an acceptable level by ensuring still acceptable results. A scheme of the process is shown in Figure 4.2. Additionally, the aerodynamic analysis is moved away from a normal desktop computer to a high-performance machine for parallel computation. The size of the unstructured mesh for the treatment of a supersonic business jet is in the order of 150000 grid points encompassing about 750000 tetrahedra and 120000 surface triangles. An example of the automatically generated surface mesh is shown in Figure 4.3.

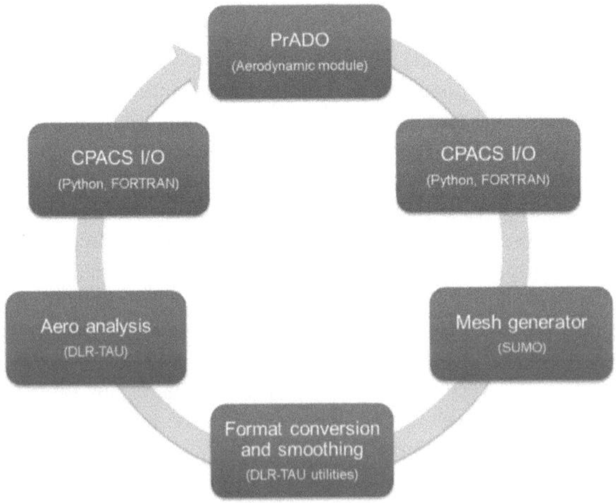


**Figure 4.2:** Mesh generation process

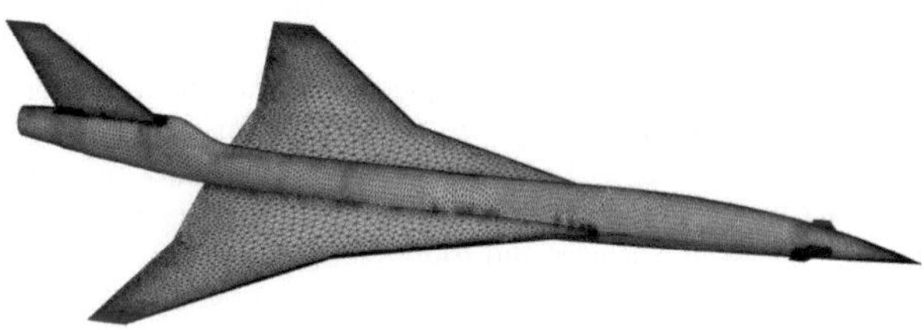

**Figure 4.3:** Surface mesh generated by SUMO and enhanced by TAU utilities. Geometry based on HISAC aircraft, cf. Deremaux (2009)

The following part of the text has been published previously by the author in Schuermann et al. (2014, 2015).The main computation process is carried out by means of a Python control script. In a first step, the required cases to be analysed are read from the CPACS file, along with the reference conditions to be used for the computation of the (non-dimensional) force and moment coefficients. The computations are then performed by looping over all required Mach numbers, angles of attack and yaw angles. During an angle of attack sweep at constant Mach number the flow field is not reinitialized with free stream values if the change in the angle of attack is smaller than a threshold value, thus saving computational time. Additionally the computational settings are varied depending on the Mach number, since convergence speed and robustness of the computation depend heavily on the flow conditions. In a post-processing step, the results (force and moment coefficients) are written into the CPACS file. Since the Euler equations do not take the viscosity of the fluid into account, an estimation of the viscous friction drag is added to the coefficients as explained in Sections 3.1.4 and C.1. Given that viscous friction drag is estimated as outlined by means of a very simple analytical method, viscous pressure drag is not calculated. This is acceptable for two reasons: Firstly, for cruise conditions, where the flow is attached and the boundary layer is very thin due to the high Reynolds number, the correction would be lower than the error margin of the viscous drag estimation. Secondly, at low speed, it would have greater impact, but it would also be the case for other effects, in particular the influence of viscosity on the vortical flow at high angles of attack. For these reasons, an estimation of viscous pressure drag is beyond the scope of this work. Details on the computation of aerodynamic loads for the structural model are given in Section 4.3.

## 4.2 Structural Model

The finite element model (FE model) comprises the aircraft's main components, such as lifting surfaces and spindle-shaped surfaces like fuselages. Modelling of the complex structural layout of aircraft is not possible within the framework of preliminary design. Therefore, some assumptions and simplifications have to be made.

Firstly, thought has to be given to the level of detail. Besides the external shape of the aircraft, its internal structure requires modelling. This inner layout consists of ribs and spars for lifting surfaces and frames for spindle shaped surfaces. Special features like stringer separation, stringer cross-sections, rivet or welding connections, their compliances and many others are not modelled in detail within the framework of numerically efficient preliminary aircraft design. Additionally forces resulting from the flow around engines are neglected. Their integration would only be possible with tremendous additional effort regarding model generation and analysis time at this stage of the development process.

Secondly, the static analysis is geometrically and physically linear. Equilibrium of the non-deformed body (rigid aircraft) and a linear material model are assumed. The finite element model that is currently available in PrADO is proposed by Österheld

(2003). It uses a single two-dimensional element for each rib/frame bay, which provides membrane stiffness only. The model allows for the creation of *monocoque* and *semimonocoque* structures. Stringers are not modelled explicitly. Their influence has been taken into accounted by a surrogate model. The model is a half model with symmetry along the fuselage's longitudinal axis.
In contrast to the model by Österheld, the finite element model proposed in is thesis is generated using 2D shell elements for representation of the skin and web structure (e.g. spar web). The elements could have either isotropic or orthotropic behaviour. Fuselage frames and the flanges/caps of spars are represented by 1D beam elements of user defined cross section. Consequently, the modelling approach is consistent in terms of degrees of freedom. The use of the mentioned element types allows for a refined mesh compared to the current model. Additionally, the model generator implements the parametric geometry description delivered by PrADO more strictly and allows to use either a half or a full model. The full model allows for the generation of more complex aircraft configurations, e.g. more than one vertical tailplane or fuselage. The surrogate stringer model by Österheld (2003) is implemented and will be explained in brief in Section 4.2.2.

### 4.2.1 Used Finite Element Types

In the above section, the use of beam elements as well as shell elements is mentioned. Even though the theory regarding these element types is to be found in the accompanying documentation to the herein used finite-element program MSC.NASTRAN® (e.g. Anon, 2011) or other pertinent literature, some aspects shall be mentioned here that help to better understand the analysis model.
One-dimensional beam elements[2] based on the Euler-Bernoulli beam theory are used in this model. This theory assumes that no deformations occur in the plane of the cross section. Furthermore, the cross section remains plane after deformation and it remains normal to the neutral axis of the beam. Consequently, transverse shear stiffness along the length of the element is neglected. The element properties are constant along the length of the element. Rod elements are used in some areas of the model to support the cabin floor. These elements are of circular cross section and carry axial load only. The element properties are constant along the length of the element.
Three types of surface elements are used in the model. An isoparametric shell element constituted of four nodes is the preferred element. The element formulation provides membrane and bending stiffness to the model. The thickness of the element is constant across its surface, whereas its properties could be either isotropic or orthotropic. Each node has six degrees of freedom, albeit the element does not provide *"direct elastic stiffness to the sixth degree of freedom, i.e., the rotation about the normal to the surface of the element"*(Anon, 2011). In some cases, mesh generation is eased by the application of an isoparametric shell element constituted of three

[2] In MSC.NASTRAN® the name for the element type used herein is CBAR, which may be confusing having the differences between bars and beams in mind as explained in Section 3.2.4.

nodes. The aforementioned statements stay valid for this element type. Stresses are taken from the bottom and top surfaces of elements with isotropic properties, while stresses of elements of orthotropic properties are evaluated for each layer. Shear panels are used in the model as floor panels. Shear panels carry the in-plane shear, whereas beams that constitute the floor grid carry extensional and bending loads. The use of shell elements for this purpose would result in an overly stiff model.

### 4.2.2 A Surrogate Stringer Model

Stringers are stiffening elements (in longitudinal direction) that increase the stiffness of the aircraft's skin and thus reduce the risk of buckling. They contribute directly to the deformation behaviour of the structure and exist in fuselage and lifting surface designs. For the purpose of the surrogate model, it is assumed that the stringers are evenly smeared over the domain. Consequently, each skin element consists of a minimum of two layers, one of which is the stringer layer which has orthotropic properties (also referred to as orthotropic layer). The stiffness of the orthotropic layer is defined along the longitudinal axis of the stringer and normal and shear stiffness are neglected. This methodology treats the skin stiffness and the stringer stiffness separately, which according to Österheld (2003), ensures that the combined stiffness of a component is represented realistically. The skin thickness and the thickness of the orthotropic layer are related to each other by the use of the factor $c_{ortho}$. Equations 4.2 and 4.3 show the relation. When applying this approach to the model described in Section 4.2, one has to keep in mind that secondary bending occurs when the element is loaded in tension due to the geometric eccentricity. This results in increased stress levels.

$$t_{str} = c_{ortho} \cdot t_{total} \tag{4.2}$$

$$t_{skin} = (1 - c_{ortho}) \cdot t_{total} \tag{4.3}$$

Here, $t_{str}$ is the thickness of the stringer layer, $t_{skin}$ denotes the thickness of the skin layer and $t_{total}$ is the total thickness of the element. In the current implementation of the code, the factor $c_{ortho}$ is based on engineering judgement, and results obtained by Österheld (2003) and Rieke (2013). They are defined by the user for each subcomponent of the aircraft. The chosen values are based on results obtained by Österheld (2003) and Rieke (2013). It is important to note that within the sizing process, only the membrane stiffness of the skin is modified by adapting the skin thickness and the thickness of the orthotropic layer is determined using the above-mentioned relations. The relations are shown in Figure 4.4

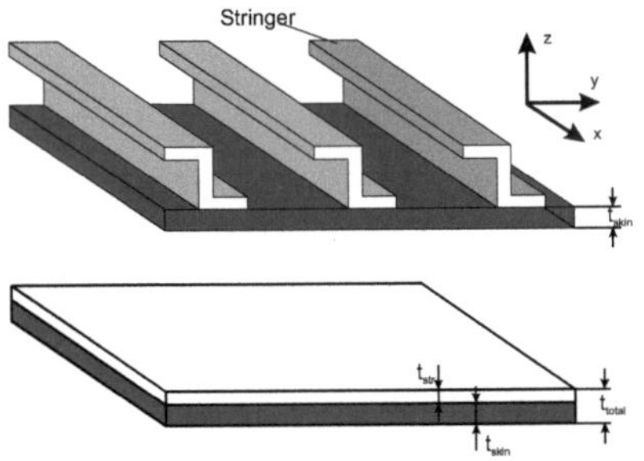


**Figure 4.4:** Surrogate stringer model: flat plate with three stringers (top), surrogate model based on Österheld (2003) (bottom)

## 4.3 Loads, Load Cases and Boundary Conditions

A differentiation between load-case dependent and load-case independent masses is made in the model[3]. Load-case independent masses are masses that are constant for all load cases, such as engine mass, masses of the secondary structure like flaps, cabin interior, lights, systems and many others. These masses result in inertia loads when exposed to an acceleration. The data and a detailed distribution are taken from the PrADO database. In the FE model these masses are represented by concentrated masses (point masses). Interpolation elements attach these masses to the structure. According to Anon (2011, p. 327) the elements used, do not add additional stiffness to the model.

The mass distribution for payload is taken from the PrADO database at discrete points, e.g. at the location of seats or containers. The masses are converted into forces. The resulting forces are introduced into the model using the above-mentioned interpolation elements. It is evident, that loads based on payload are load-case dependent.

Fuel masses are modelled as forces that act on the centre of gravity of the fluid. The fuel mass and its centre of gravity are integrated into the model using the above-mentioned interpolation elements. Fuel tanks that are integrated into lifting surfaces are connected to the bottom surface of the structure. Fuel tanks in the fuselage structure are connected to all nodes below the centre of gravity of the fluid.

The aerodynamic forces are taken from an analysis by TAU (Schwamborn et al. (2006)), which solves the inviscid Euler equations for a given target lift coefficient. The load cases are provided by PrADO and developed by computing the V-n Diagram (cf. Niu, 1999) from which the target lift coefficient $C_{lift}$ is calculated based

[3]At this stage, it seems necessary to illustrate the definition of mass, weight and force. Generally speaking, a *mass* is a quantity or aggregate of matter. *Weight* on the other hand, is a force resulting from applying a gravitational acceleration on a mass. Finally a *force* is an influence exerted on an object that causes it to change its state (Anon, 2014a).

on equation 4.4, in which $m$ is the mass of the aircraft for the given load case, $n_z$ is the load factor derived from the V-n Diagram, $g$ represents the gravitational acceleration and $Ma$ is the flight Mach number. $\rho$ and $a$ are the density and speed of sound, respectively.

$$C_{lift} = \frac{2 \cdot n_z \cdot m \cdot g}{\rho \cdot (a \cdot Ma)^2 \cdot A_{ref}} \tag{4.4}$$

The resulting aerodynamic forces are transferred onto the structural analysis model by *ifls*, while a rigid body is assumed. *ifls* is a tool that provides coupling techniques for fluid-structure interaction (Haupt et al., 2005). The attachment of leading- and trailing-edge device masses is shown in Figure 4.5. Further examples for the introduction of payload, fuel masses and aerodynamic forces are depicted in Figures A.1(a) and A.1 of Appendix A.

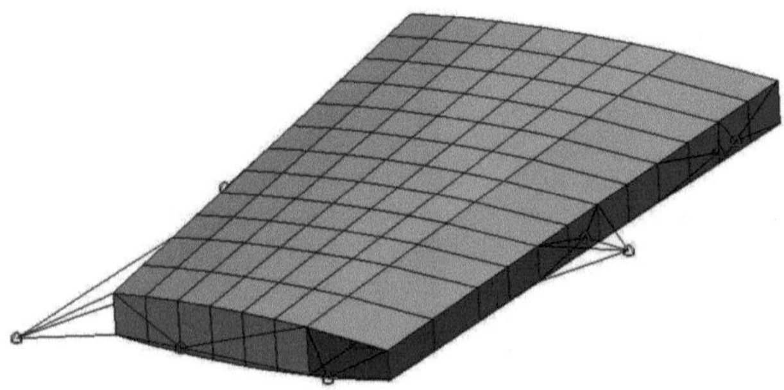

**Figure 4.5:** Trailing and leading edge point masses (red) in the finite element model

The boundary conditions for flight cases are imposed on the wing root and on the wing's symmetry plane. This is shown in Figure 4.6(a) where the red markers indicate the positions at the wing root. For ground cases the conditions are imposed on the landing gear struts. This is shown in Figure 4.6(b) where again the makers indicate the positions. For flight cases, the selected displacement conditions constrain the translatory motion, while rotation is constrained by a combination of the former. For ground cases the translatory motion and rotations are restrained, thus the system is statically overdetermined.

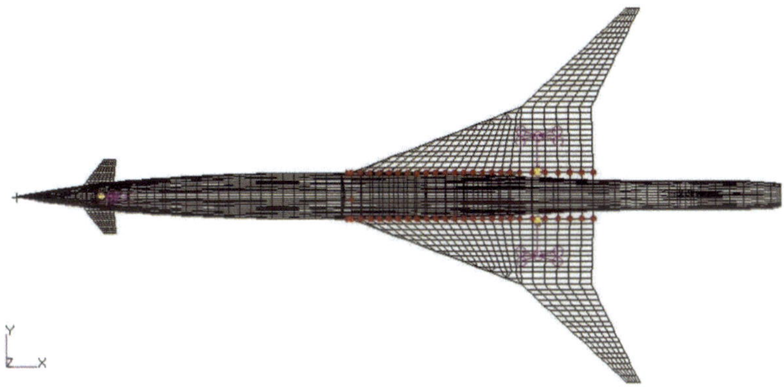


(a) Flight case - top view

(b) Ground case - side view

**Figure 4.6:** FEM boundary conditions

# 5 Verification

The purpose of this chapter is to provide verification examples for aspects discussed in Chapters 3 and 4. Therefore, aerodynamic data is presented in Section 5.1 and the structural sizing process is judged against an example from literature in Section 5.2. Eventually, the tool chain is verified in its entirety in Section 5.3 where global aircraft data is compared with a reference.

## 5.1 Verification of Aerodynamics

This section is subdivided into two parts. The first presents aerodynamic data for verification purposes, whereas the second is a brief comment on numerical issues.

### Verification of Aerodynamic Coefficients

The aerodynamic procedure can hardly be verified or even validated due to the lack of reference data, as e.g. from higher-order numerical simulation or existing supersonic aircraft. The configuration presented in Chapter 6 is solely created to provide information on temperature behaviour and due to modelling issues not suited to provide data for this section. However, a verification based on the qualitative behaviour of the data is performed giving certainty that the process works correctly, even though the numerical values are not looked at. For this reason various data in dependence of the Mach number is plotted in Figure 5.2(a) and compared to literature and a selected distribution of the pressure coefficient, $C_p$ , is depicted in Figure A.2. The aircraft configuration given in Figure 5.1 serves as input for the verification process. A detailed description of this configuration is given in Section 7.3 along with additional data. It serves a basic design for further analysis at that later stage.
It can be observed in Figure A.2 that the bottom surface shows negative coefficients whereas the top surface shows positive values. Furthermore, leading-edge suction can be observed. A large part of the bottom surface shows a homogeneous $C_p$ distribution, whereas the distribution of the top surface is more scattered. The pressure coefficient reaches its maximum at the nose of the fuselage. Imperfections of the fuselage's geometry can be seen in various regions. However, the observations fulfil the general expectations for the given flight condition.

*Total drag*:
In Figure 5.2(a) total drag, $C_{drag}$, increases from $Ma = 0.5$ until the sound barrier and reaches its maximum slightly after $Ma = 1.0$ before it decreases. In general, drag in the supersonic regime is higher than in subsonic flow. This corresponds

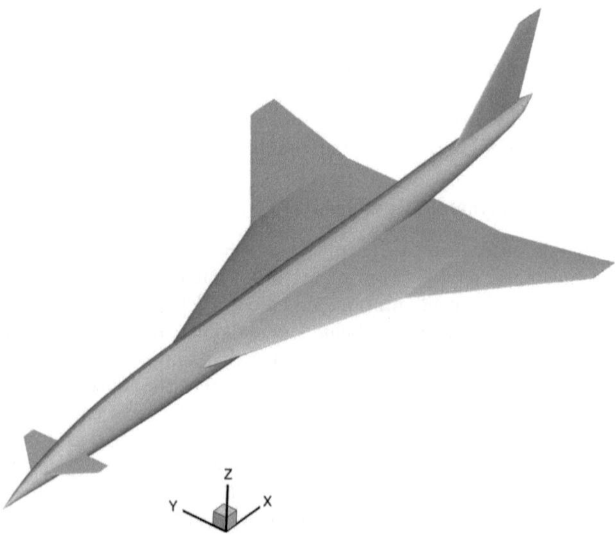


**Figure 5.1:** Verification of aerodynamics: geometry input - PrADO representation

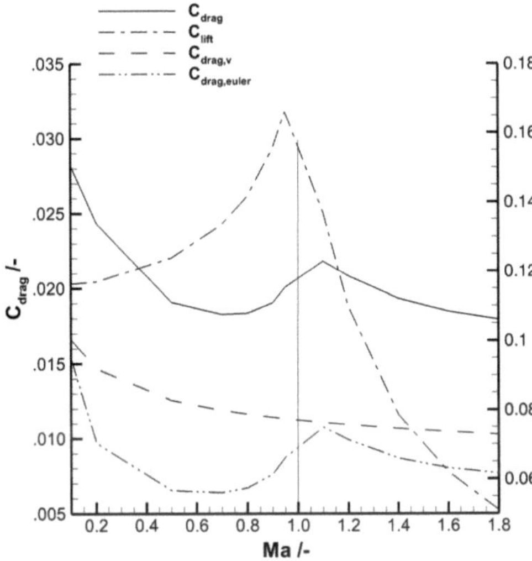


(a) Aerodynamic coefficients

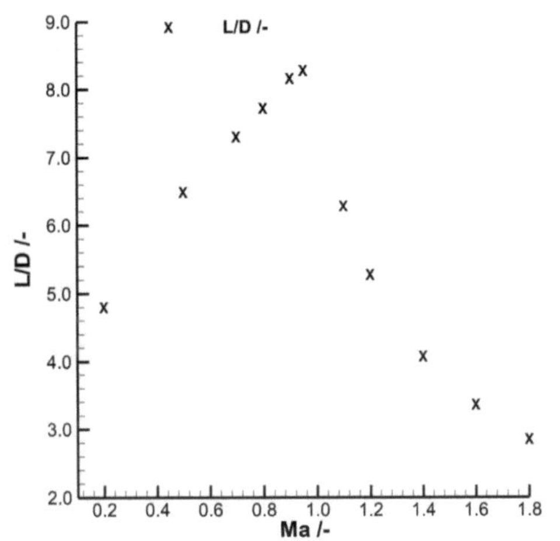


(b) Lift-to-drag ratio ,$L/D$

**Figure 5.2:** Verification of aerodynamics: results ($Altitude = 18km$ $\alpha = 0deg$)

with what is shown in literature (Heinze (2007)). Literature (Heinze (2007)) suggests that the gradient from low Mach numbers up to about $Ma = 0.5$ is negative, which proves true for the data presented here. The frictionless solution ($C_{drag,euler}$), hence the data obtained from TAU before viscous drag estimation, follows the same trend, thus indicating the origin of the behaviour as being the frictionless TAU solution. The difference between total drag and frictionless drag is in this case viscous drag, $C_{drag,v}$. The procedure of its computation is explained in Section 3.1.4. The slope of $C_{drag,v}$ compares well with Heinze (2007). It can nevertheless be stated that the overall behaviour of the drag coefficient in the region of the relevant Mach number matches the expectations.

*Lift coefficient and lift-to-drag ratio*:
In Figure 5.2(a), the slope of the lift coefficient compares well with what is shown in literature (Heinze (2007)). However, it has to be noted that the maximum lies right below $Ma = 1.0$, where it would be expected to be above. The reason in this case could be that no values have been computed for the aerodynamic performance map between $Ma = 0.95$ and $Ma = 1.1$, since automated CFD analysis in this flow regime proofed to be very complex and time consuming. Therefore, it is expected that the maximum value would be shifted to a Mach number above $Ma = 1.0$, should values between $Ma = 0.95$ and $Ma = 1.1$ be derived. The overall behaviour meets the expectations. For reasons of completeness the lift-to-drag ratio, $L/D$, is given in Figure 5.2(b). It can be observed that the values in the supersonic regime ($Ma > 1.2$) are well below those of the subsonic regime. This and the general slope meet the expectations.

### Remarks on Numerical Behaviour and Settings

The solution is computed using the TAU multigrid scheme 2v. This scheme converges slower than other schemes, but is more robust when computing flow with shocks. The upwind schemes used are *van Leer* and *AUSMDV*. The former is used to initialize a solution, whereas the later drives the solution towards convergence. The scheme *AUSMDV* resolves shocks better than *van Leer* but is less stable. By starting *AUSMDV*, from a given solution numerical stability is enhanced. However, the settings are a confluence of information given in the TAU user guide and experience. Detailed information is given in the TAU user guide (Anon (2014b)) and other software-related documentation.
Figure 5.3(a) shows the convergence behaviour of the TAU main run[1] for $Ma = 1.6$ and $\alpha = 0deg$. Convergence of the residual, the lift coefficient $C_{lift}$ and the drag coefficient $C_{drag}$ is reached by little more than 3000 iterations. The scenario is different when computing subsonic flow. Figure 5.3(b) shows the convergence behaviour of the main run for $Ma = 0.5$ and $\alpha = 0deg$. The main run starts after 1500 iterations and more than 16000 iterations are required to reach convergence.

[1]This is why the abscissa starts at 1500 iterations.

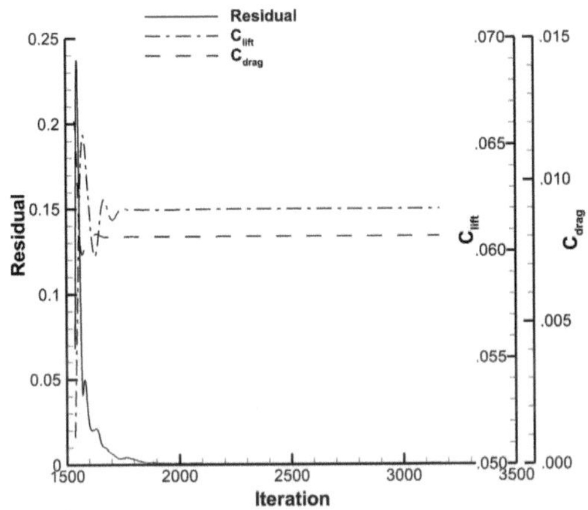


(a) Convergence for $Ma = 1.6$ and $\alpha = 0deg$

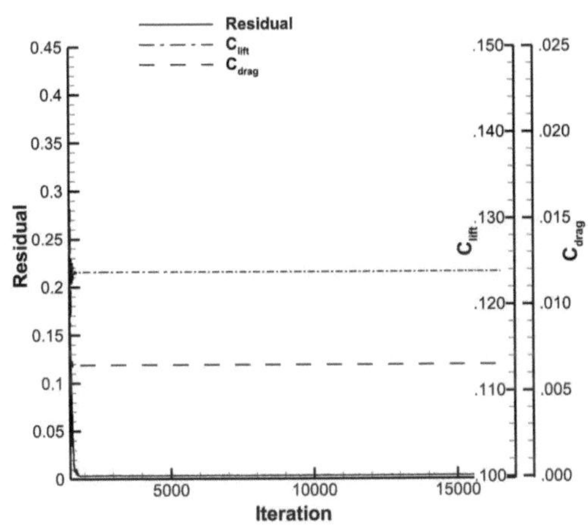


(b) Convergence for $Ma = 0.5$ and $\alpha = 0deg$

**Figure 5.3:** Verification of aerodynamics: convergence behaviour

A similar behaviour is observed for the analysis in the transonic regime. The differences are due to the different characteristics of the flow. Subsonic flow is parabolic whereas supersonic flow is of hyperbolic nature which in general takes more time to being solved.

## 5.2 Verification of the Structural Sizing Module

The verification of the structural sizing process, hence the structural sizing module (SSM), is based upon analysis of a conventional wingbox made out of aluminium alloy. The underlying analysis model is the wing of the generic DLR-F11 aircraft as presented in Hürlimann et al. (2011, 2012). As the original FEM input data of this study has been made available by the authors of Hürlimann et al. (2012), no re-modelling was needed. Thus, errors or misinterpretation of the data is avoided, resulting in the desired pure verification of the sizing algorithm.

### Model Description

The span of the aircraft is 53.4 meters with a maximum take-off mass[2] ($m_{MTOM}$) of $230300kg$. The maximum fuel capacity is $130400kg$. Loads are introduced as SMT (Shear Moment Torsion) cut loads. The wing in its finite element representation is shown in Figure 5.4. The model is subject to eight load cases of which the data is shown in Table B.3 of the Appendix. Besides the number of the load case, information on the load case type, the load factor, $n$, the altitude, *Alt*, the flight Mach number, *Ma*, and the aircraft mass are given.

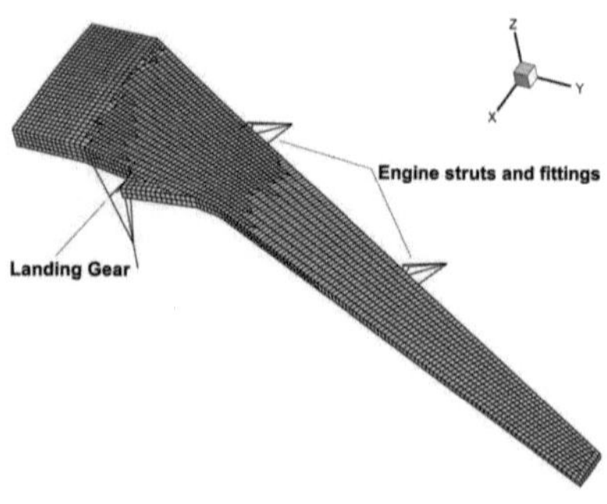


**Figure 5.4:** DLR-F11 wing: structural model

[2]The expression "maximum take-off weight ($m_{MTOW}$)" is often used synonymously.

Minimum gage constraints have been imposed on the model. They are summarised in Table B.4 and material data is given in Table B.5 of the Appendix. For further details references Hürlimann et al. (2011, 2012) are suggested.

#### Results

In a first step, the Structural Sizing Module (SSM) is run with default settings (cf. Table B.2, Appendix B). The design variables are sized based on stress and yield strength. The computed values are compared with data from table 2 of Hürlimann et al. (2012). The results are shown in Table 5.1. It should be noted that the SSM provides one mass value for all spars, whereas the reference provides data separately for each spar.

**Table 5.1:** DLR-F11 wing: comparison of results for SSM verification

| *Group name* | *Reference mass /kg* | *SSM mass /kg* | $\Delta mass$ */kg* | $\Delta mass$ */ %* |
|---|---|---|---|---|
| Ribs | 1750.0 | 1749.92 | 0.08 | 0.00 |
| Front Spar | 334.1 | | | |
| Mid Spar | 235.9 | 984.76 | 2.74 | -0.28 |
| Rear Spar | 417.5 | | | |
| Upper Skin | 4357.9 | 4371.99 | 14.09 | 0.32 |
| Lower Skin | 4936.2 | 5044.84 | 108.64 | 2.20 |
| **Total** | **12031.6** | **12151.51** | **119.90** | **1.00** |

The data compares well with the reference. The comparison of Figures 5.5 and 5.6 shows that the thickness distribution is qualitatively comparable. The former figure is the result obtained by SSM, whereas the later results from Hürlimann et al. (2012), Figure 13, top. The quantitative differences are acceptable. However, it is to be noted that the maximum values obtained with the SSM ($t_{max} \approx 37mm$) are higher than those obtained by Hürlimann et al. ($t_{max} \approx 30mm$) and, therefore, the scale in the figures differs slightly. Convergence behaviour is presented in Figure 5.7 where it can be seen that both convergence criteria are satisfied after twelve iterations. Convergence of mass is reached after seven iterations, whereas convergence of the maximum failure index is reached at iteration twelve. Note that the ordinate starts at a value lower than 0.0 to enhance visibility of the data.

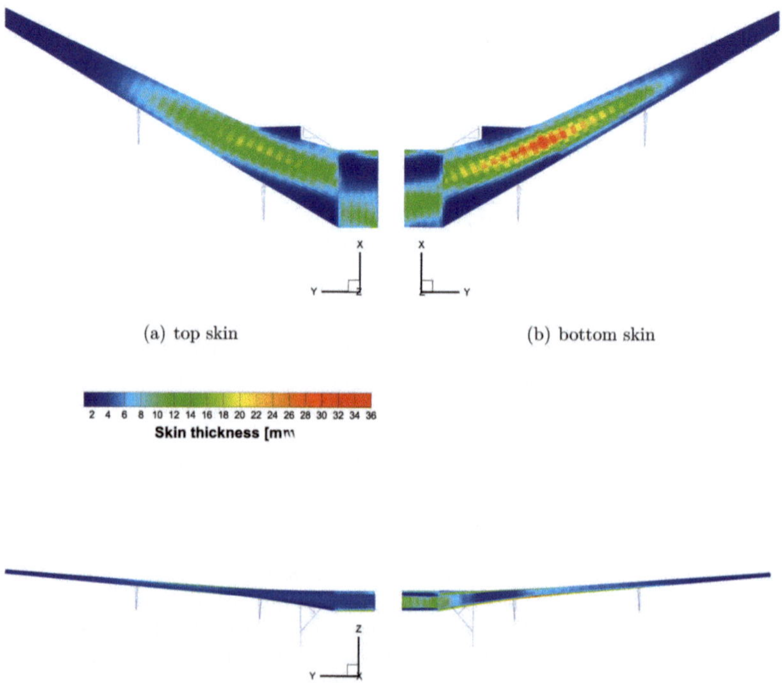


(a) top skin

(b) bottom skin

(c) front spar

(d) rear spar

**Figure 5.5:** DLR-F11 wing: skin thickness obtained with SSM

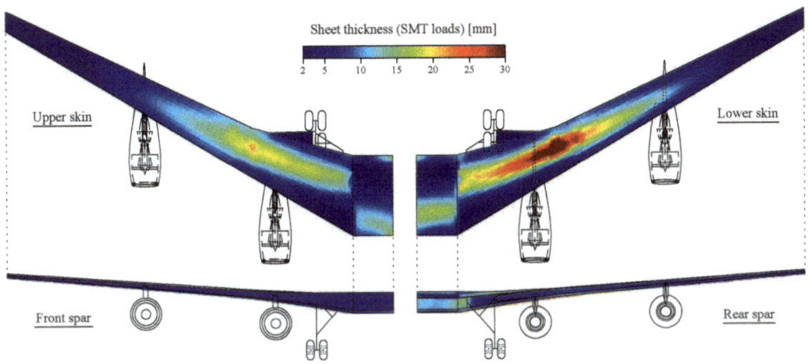


**Figure 5.6:** DLR-F11 wing: skin thickness obtained by Hürlimann et al. (2012), courtesy of F. Hürlimann

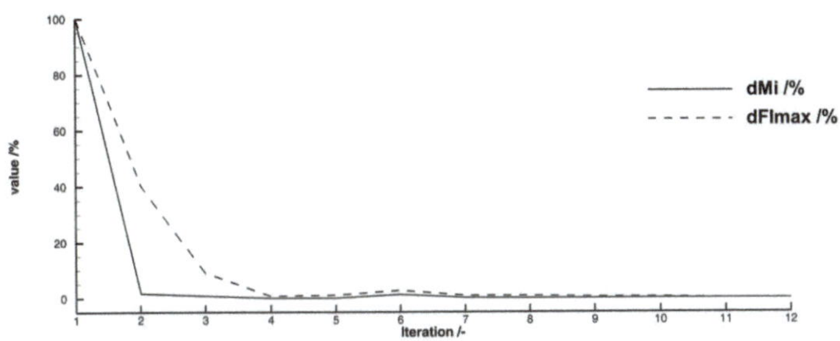


**Figure 5.7:** DLR-F11 wing: convergence of structural sizing for parameters $dFI_{max}$ and $dMi$ with default settings

## 5.3 Verification of the Tool Chain

The intent of this section is to verify the integration of methods developed and proposed in Chapters 3 and 4 into the PrADO tool chain. Therefore, the low noise aircraft configuration of the European HISAC (High Speed Aircraft) project is used as a reference (cf. Deremaux (2009)). Part of this data has been previously presented by the author in Schuermann et al. (2013, 2014) and Schuermann et al. (2015).

### Model Description

*"The aircraft is designed to carry a payload of 726 kg, which equals eight passengers and their luggage over a distance of 4000 nautical miles (7408 kilometres) at Mach 1.6. The primary structure of the aircraft consists of carbon fibre reinforced plastics (CFRP) for the cover sheets of fuselage and lifting surfaces and aluminium alloy for other structural elements. The aircraft comprises six fuel tanks in the inner part of the wing and two fuel tanks in the fuselage. Two of the three engines are integrated in nacelles under the wing, whereas one is integrated into the rear fuselage. The aircraft has one main passenger door and an emergency exit. An additional cargo door is integrated on the starboard side right after the passenger cabin and before the rear pressure bulkhead"* (Schuermann et al., 2015). Figure 5.8 shows the PrADO representation of the aircraft. Geometrical data is presented in Table 5.2.

**Table 5.2:** HISAC aircraft: geometrical data

| *Parameter* | *Reference value, Deremaux (2009)* | Unit |
|---|---|---|
| Aircraft length, $l_{AC}$ | 36.80 | $m$ |
| Wing area, $A_W$ | 150.0 | $m^2$ |
| Wing span, $s$ | 18.50 | $m$ |
| Leading-edge sweep angles, $\varphi$ | 72.5, 52.0 | $deg$ |
| Dihedral angle, $\Gamma$ | 0. | $deg$ |
| Number of engines, $n_E$ | 3 | – |

### Results

*"The goal of the analysis is to determine the Maximum Take-off Mass ($m_{MTOM}$) for a given payload and range while keeping the shape of the aircraft fixed. The above-mentioned mission requirement, the aircraft's geometry and a $m_{MTOM}$ of 51100 kg are the starting point for the iterative analysis with PrADO. The load cases taken into account for the design are presented in Table B.6 of the Appendix. It has to be noted that the values presented are valid for the final design iteration only. Data for previous steps is calculated as needed and differs form the values in Table B.6, since aircraft mass and performance data change from iteration step to iteration step until convergence is achieved"* (Schuermann et al., 2015). Table 5.3 presents

**Figure 5.8:** HISAC aircraft: PrADO model

the results of the analysis and compares them with data from Deremaux (2009). The Maximum Take-off Mass resulting from PrADO is 1.94 percent less than stated in the reference. The maximum fuel mass and the empty mass[3] compare well against the reference data. The dry runway length is questionable and might result from an overestimation of high-lift device derivatives. However, the runway length does not influence the design, since it is only a boundary condition. Two values need further explanation. The total static thrust for the HISAC configuration is derived under the assumption that the maximum net take-off thrust is 78% of static thrust. Reference Deremaux (2009) gives 220kN for maximum net take-off thrust. The PrADO value for the Empty Weight ($m_{EW}$) is the Manufacturers Empty Weight ($m_{MEW}$), since the definition of the Empty Weight used by Deremaux (2009) is not detailed.

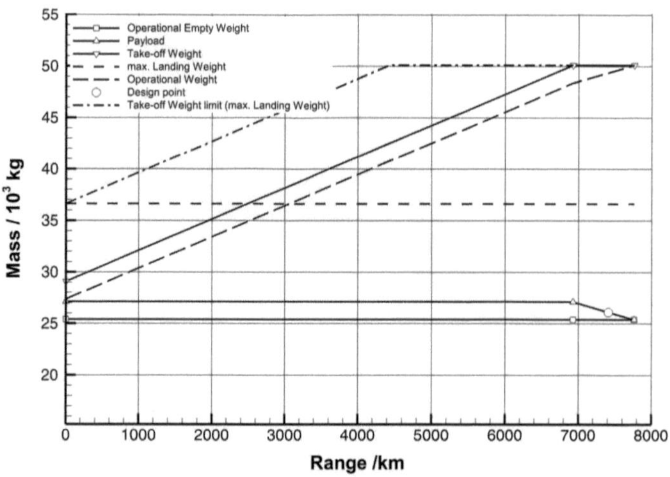


**Figure 5.9:** HISAC aircraft: mass-range chart based on PrADO analysis

A mass-range chart is depicted in Figure 5.9. One can observe in this chart that the range for the flight with maximum fuel and the range for a ferry flight are close together. Based on data from Table 5.3, it can be shown that for a flight with maximum fuel the resultant payload reduces to 5 kg, which can be considered. Hence, a flight with maximum fuel results in the same configuration than a ferry flight.

[3]The expression "empty weight" is often used synonymously.

**Table 5.3:** HISAC aircraft: comparison of PrADO results with data from Deremaux (2009) and additional PrADO data

| *Parameter* | *Ref. value* | *PrADO value* | *Unit* | *Diff. /%* | *Remark* |
|---|---|---|---|---|---|
| $m_{MTOM}$ | 51100. | 50107. | $kg$ | -1.94 | Max. take-off mass |
| $m_{EW}$ | 23100. | 24397. | $kg$ | 5.61 | Empty mass (weight) |
| $m_{OEW}$ | - | 25382. | $kg$ | - | Operational empty mass (weight) |
| $m_{fuel,max}$ | 26900. | 24720. | $kg$ | -8.10 | Max. fuel mass |
| $W_{S,max}$ | - | 334.024 | $\frac{kg}{m^2}$ | - | Max. wing loading |
| $S_0$ | 282.05 | 283.40 | $kN$ | 0.48 | Total static thrust |
| $ALT_{begin,cruise,D}$ | - | 14.341 | $km$ | - | Altitude at the begin of cruise at the design point |
| $ALT_{end,cruise,D}$ | - | 17.319 | $km$ | - | Altitude at the end of cruise at the design point |
| $L/D_{begin,cruise,D}$ | - | 6.82 | $m$ | - | Lift-to-drag ratio at the begin of cruise at the design point |
| $V_{AT}$ | - | 193., (104.) | $\frac{km}{h}, (kts)$ | - | Speed at threshold |
| $l_{RWY,T}$ | - | 908. | $m$ | - | Runway length at take-off |
| $l_{RWY,LD}$ | 1700. | 1095. | $m$ | -35.6 | Runway length at landing |
| DOC | - | 0.38756 | $\frac{\text{EURO}}{SKM}$ | | Direct operating costs |

# 6 Remarks on Temperature Effects in Supersonic Flight

In the last century, research was conducted in order to analyse the behaviour of structures exposed to high temperatures due to supersonic flight. With the dismissal of Concorde from service, the interest in this topic decreased within the scientific community. Supersonic flight is only conducted by the military, hence should there be information on the subject, it will most likely not be publicly available. According to Torenbeek et al. (2004), it is not required to consider temperature effects in aircraft conceptual and preliminary design for velocities lower than $Ma = 2.0$. More recently Huda & Edi (2013) published results regarding the selection of composite materials for a supersonic aircraft. However, the authors do not comment on how the used temperatures were obtained. A master thesis by Bekemeyer (2014), initiated and supervised by the author, aims at being the starting point to close that gap. Results are worth to be shown here to provide solid ground for future discussion and work on the subject, albeit the configurations that will be treated later in this thesis cruise at velocities lower than $Ma = 2.0$. In the subsequent sections, the used model will be explained and results summarized. A short theoretical introduction to handbook methods and the used CFD scheme is given. A detailed theoretical derivation is neglected; in this respect, reference is made to Bekemeyer (2014) and the relevant literature.

## 6.1 Theory

In the subsequent sections, a brief introduction to temperature prediction methods for temperature due to compressibility effects and skin friction is given.

### 6.1.1 A Handbook Method for Temperature Prediction

The increase in temperature $\Delta T$ resulting from compressibility effects and friction for supersonic flow on the aircraft's outer surface is estimated by the use of boundary layer theory (cf. Schlichting (1982); White (2005)). The theory is valid for an ideal gas, which is a fine assumption for air upto $Ma = 6$. Furthermore, the theory is only valid for a laminar boundary layer and a flat plate, which is why the subsequent formulae are an approximation. According to Schlichting & Truckenbrodt (2001), the temperature increase at the stagnation point due to compressibility effects can be expressed in equation 6.1, whereas $V_\infty$ denotes the freestream velocity and $c_p$ is the specific heat at constant pressure.

$$\Delta T = \frac{V_\infty^2}{2c_p} \tag{6.1}$$

Replacing $V_\infty$ by the Mach number $Ma$ and assuming the heat capacity ratio for air as being $\kappa = 1.405$, results in equation 6.2 to calculate a temperature increase due to compressibility at the stagnation point. $T_\infty$ denotes the freestream temperature.

$$\Delta T = 0.2 Ma_\infty^2 \cdot T_\infty \tag{6.2}$$

Similar to the approach above, Schlichting & Truckenbrodt (2001) propose equation 6.3 to calculate the temperature increase due to friction for a turbulent boundary layer.

$$\Delta T = 0.175 Ma_\infty^2 \cdot T_\infty \tag{6.3}$$

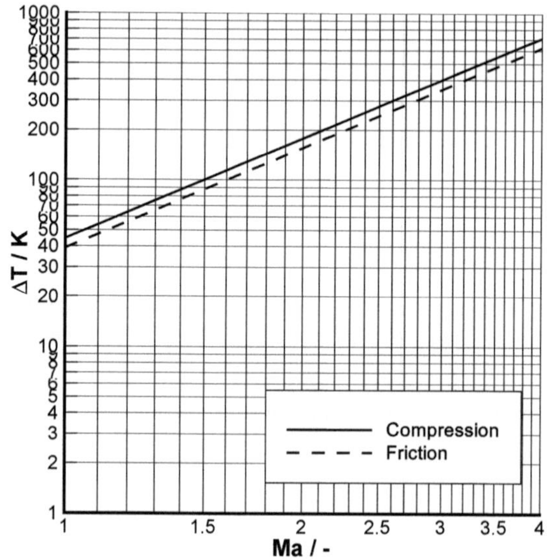


**Figure 6.1:** Temperature analysis: temperature increase due to compressibility effects and skin friction, equations 6.2 and 6.3 (logarithmic scale)

The results for a temperature increase due to compressibility and friction are shown in Figure 6.1. A different, but well-known representation of the formulae is to divide equations 6.2 and 6.3 by $T_\infty$, hence removing their altitude dependency. This is achieved in equations the 6.4 and 6.5. The validity of this assumption will be briefly outlined in Section 6.3.2.

$$\frac{\Delta T}{T_\infty} = 0.2Ma_\infty^2 \tag{6.4}$$

$$\frac{\Delta T}{T_\infty} = 0.175Ma_\infty^2 \tag{6.5}$$

### 6.1.2 CFD-Analysis for Temperature Prediction

In contrast to handbook methods, where $\Delta T$ is calculated at the stagnation point, CFD analysis can be used to predict the temperature field on the aircraft's surface. For the work presented here, the DLR-Tau code is used for this type of analysis. The Reynolds-averaged Navier-Stokes equations are solved to obtain the temperature field on the surface of the aircraft resulting from compressibility and viscous effects. This set of equations requires the modelling of turbulence, which is ensured by Menter's SST turbulence model that is part of the Tau code. This model is a two-equation eddy-viscosity model, which has the advantage that the turbulent kinetic energy is directly modelled and, therefore, the influence of turbulence on the temperature field does not need to be modelled separately.
The derivations of the Reynolds-averaged Navier-Stokes equations and of the turbulence model are available in literature and are therefore not detailed here. If further information is desired, the well-disposed reader is advised to consult Menter (1994) and Wilcox (1998) for turbulence modelling and White (2005) and Schlichting (1982) for the Reynolds-averaged Navier-Stokes equations.

## 6.2 Model Description

The following sections provide data on aircraft geometry and information on the CFD analysis model.

### 6.2.1 Geometry

Two changes have been made compared to the HISAC configuration in Chapter 5.3. The aircraft does have a conventional empennage instead of a canard and the configuration comprises only two engines. The geometry of the fuselage has been taken from a study conducted in Schuermann et al. (2014) and is based on the European HISAC project (cf. Deremaux (2009)). The design of this aircraft originates from a not further pursued configuration for eight passengers. The aircraft is shown in Figure 6.2. The wing is build from NACA 25XX aerofoils and varies in its thickness in spanwise direction. Symmetric NACA 00XX aerofoils compose the

**Figure 6.2:** Temperature analysis: aircraft configuration

horizontal and vertical tailplane. Their thickness varies in spanwise direction. The decision to use NACA aerofoils for this study is based on the fact that their data is publicly available. Tables B.7, B.8 and B.9 provide detailed information on the components.

### 6.2.2 CFD Model

The CFD analysis model is a half model, since the flow considered is symmetric with respect to the x-z plane. Figure A.3 shows certain aspects of the mesh that are discussed in the following text.

First, a structured quadrilateral mesh is created on the surface. This mesh is needed to capture viscous effects in the boundary layer. It is composed out of 100 cells of increasing thickness resulting in a total thickness of about 1 metre for the boundary layer. The mesh at the wing tip, the tip of the HTP (horizontal tail plane) and VTP (vertical tail plane), as well as the wake are modelled by an unstructured mesh, since this enhances convergence of the overall model. However, the unstructured mesh makes the solution in these regions more mesh-dependent. An unstructured mesh is used to model the farfield, which allows reducing the number of cells with increasing distance from the aircraft, hence resulting in a reduction of analysis time. Furthermore the geometry of the farfield is that of a cone with just a few cells in front of the

**Table 6.1:** Temperature analysis: combinations of altitude and Mach number

| | Ma /- | | | | | | | |
|---|---|---|---|---|---|---|---|---|
| alt /km | 1.4 | 1.6 | 1.8 | 2.0 | 2.2 | 2.5 | 2.75 | 3.0 |
| 7.5 | - | - | - | - | X | - | - | - |
| 10.0 | X | X | X | X | X | X | X | X |
| 15.0 | - | - | - | - | X | - | - | - |
| 20.0 | X | X | X | X | X | X | X | X |

aircraft. This is possible for supersonic flow because of its hyperbolic characteristic, which means that information is only carried in the upstream direction. The opening angle of the cone results directly from the Mach angle. By modifying equation 4.1, the opening angle of the Mach cone is obtained through equation 6.6.

$$\mu = \arcsin \frac{1}{Ma_\infty} \tag{6.6}$$

The minimum velocity assumed for this investigation is $Ma_\infty = 1.4$ resulting in an opening angle of $\mu \approx 45$ degrees. The generated grid consists of about 26 million grid points, whereas approximately 18 million solely contribute to the boundary layer, hence the structured part of the mesh. The hyperbolic characteristic of the flow is also considered for the selection of boundary conditions set for *supersonic inflow* and *supersonic outflow*. For further information on the mesh and numerical details, the well-disposed reader is advised to consult Bekemeyer (2014).

## 6.3 Results

In this sections, the results are presented and discussed. For the work presented, various combinations of Mach number, altitude and angle of attack are analysed by CFD. The resultant data is used to estimate the influence of temperature on materials. Table 6.1 shows combinations of Mach number and altitude for an angle of attack of zero degrees. Mach numbers below $Ma = 1.4$ do not seem to be reasonable from a thermal perspective, whereas Mach numbers above $Ma = 3.0$ are neglected because they are out of scope for current SSBJ concepts. In addition, Table 6.2 shows a variation of the angle of attack and Mach number for a constant altitude of 10 kilometres. Since high thermal load is expected under cruise conditions, the angle of attack variations have been limited to values between $\alpha = -2$ and $\alpha = 6$.

### 6.3.1 Temperature Field

The results for the combination of $Ma = 2.2$, an altitude of $Alt = 10.0km$ and $\alpha = 0.0deg$ are discussed in this section. Furthermore, the data is compared with the handbook method described in Section 6.1.1 and remarks on the results obtained

**Table 6.2:** Temperature analysis: angle of attack variations

| | Ma /- | |
|---|---|---|
| $\alpha$ /deg | 1.6 | 2.2 |
| −2.0 | X | X |
| 0.0 | X | X |
| 2.0 | X | X |
| 4.0 | X | X |
| 6.0 | X | X |

by Huda & Edi (2013) are made. The temperature field is depicted in Figure 6.3. It is observed that the increase in temperature is in general between $130K$ and $225K$. Values that exceed the aforementioned range do exist, however, they are locally limited (cf. Figure A.4).

To evaluate the results properly, two slices are cut from the wing at spanwise positions of $y = 2.5m$ (Cut A, inner cut) and $y = 6.0m$ (Cut B, outer cut). The data is shown in Figure 6.4. Whilst it can be observed that the temperature distribution for the cut A is oscillating, the one for cut B shows a smoother behaviour. Both distributions have a maximum close to the leading edge, whereas the values decrease towards the trailing edge. The higher maximum temperature at cut B can be explained by the reduced wing sweep. Reducing the sweep angle results in a higher velocity perpendicular to the leading edge, hence higher temperatures. A comparison of the computed data with results obtained using the method described in Section 6.1.1 shows that the values at the stagnation point ($\Delta T \approx 215K$) are in the same order of magnitude. The same is true for the temperature increase due to friction (CFD result: $\Delta T \approx 190K$). It can be observed in Figure A.4(a) that the temperature increase at the horizontal tailplane (HTP), as well as its temperature field are comparable to that of the wing. The major difference is that the aerofoil is symmetric, which results in similar distributions for the top and bottom surface of the HTP. The interference of the vertical tailplane (VTP) with the HTP can be observed in Figure A.4(a) and Figure A.4(b). As for the wing, a major temperature increase is observed at the tip of the HTP and VTP, where an unstructured mesh is used. The huge temperature increase at the leading edge of the VTP is discussed by Bekemeyer (2014) and not detailed here. However, it can be seen that the temperature level decreases to what has been observed for the wing and HTP at about 25% of the chord of the VTP. The lower part of the VTP shows interference with the HTP (cf. Figure A.4). The temperature field of the fuselage is influenced by its curvature. A reduction in diameter results in an increase in temperature. Imperfections of the fuselage's geometry can be seen in Figure A.4(c). Where fuselage and lifting surfaces adjoin each other the influence of shocks at the leading edge of the lifting surfaces on the fuselage can be observed (cf. Figure 6.3). A comparison of Figure 6.3 with data published by Huda & Edi (2013) shows that the qualitative

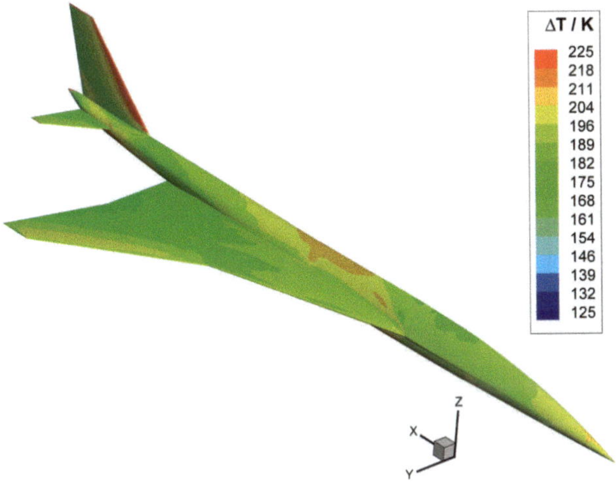


(a) top surface

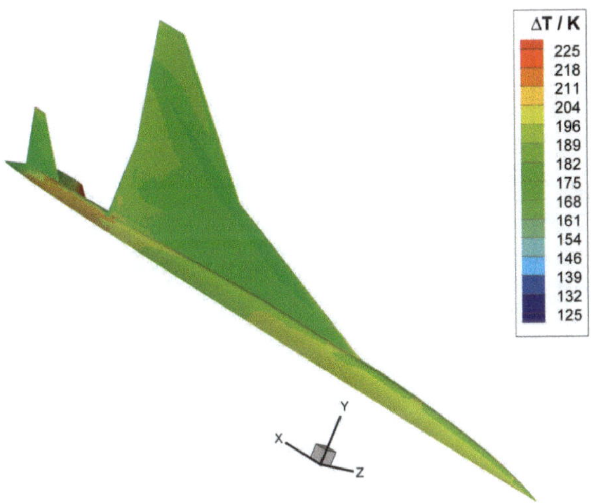


(b) bottom surface

**Figure 6.3:** Temperature analysis: temperature field obtained with CFD for $Mach = 2.2$, $altiute = 10km$ and $\alpha = 0.0deg$

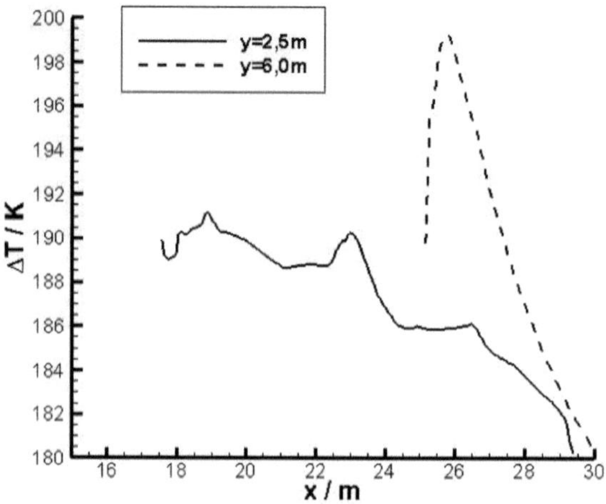


(a) top surface

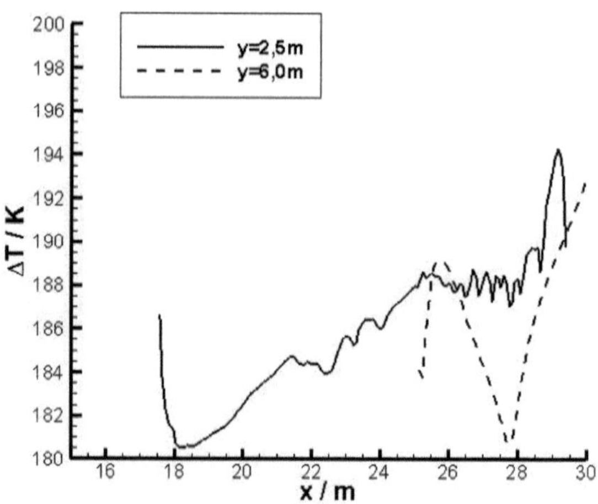


(b) bottom surface

**Figure 6.4:** Temperature analysis: results at wing stations $y = 2.5m$ (Cut A) and $y = 6.0m$ (Cut B) for $Mach = 2.2$, $altitude = 10km$ and $\alpha = 0.0deg$

behaviour of the temperature field is similar however, the maximum values in the publication are lower. A difference regarding the distribution can be observed for the wing. Albeit a huge wing sweep, the plot by Huda & Edi shows lines of equal temperature that are parallel to the leading edge, whereas Figure 6.3 suggests a more diffuse behaviour. In contrast to the results of this work, the stagnation point shown by Huda & Edi does not move towards the upper surface. However, the decrease in temperature of $\Delta T = 10K$ is common to both investigations. At this point, it is important to know that neither an altitude nor a Reynolds number is given in the publication, which makes a detailed comparison difficult.

### 6.3.2 Mach Number, Altitude and Angle of Attack Dependency of Temperature

In this section information on the dependency of the temperature from Mach number, altitude and angle of attack is given. Control points are established at the surface of the model to analyse the temperature changes due to the Mach number. Six such points are distributed over the wing's surface as well as another six over the fuselage's surface. The values measured at these control points indicate the temperature increase due to friction. Two additional control points are introduced at the stagnation point of the wing. They indicate a temperature increase due to compression. The control points are depicted in Figure A.5. To compare these values with data from handbook methods, the average of each group of values is taken. The result is shown in Figure 6.5, where both axis are of logarithmic scale. Altitude and angle of attack are constant. Based on this representation, the temperature increase obtained from CFD shows to be linear and compares well with handbook data (cf. Section 6.1.1). However, at Mach numbers higher than $Ma = 2.2$ the scatter between values increases, indicating that at higher Mach numbers the formulas should only be used with tremendous caution. The tip of the fuselage is not taken into consideration since the data does not seem to be meaningful when compared to handbook data. For an investigation into this issue, a refined mesh focussing on this region is suggested but out of scope for this work.

It was stated in Section 6.1.1 that the ratio of temperature increase and farfield temperature, $\frac{\Delta T}{T_\infty}$ is often assumed to be constant for a given Mach number, hence is independent from altitude. This assumption has been verified by the computation of data for the combinations of Mach number and altitude as detailed at the beginning of Section 6.3 in Table 6.1. The results in Table 6.3 show that differences between the two altitudes are in the range of 0.069% and 2.769% and that their magnitude is independent of the Mach number. The values obtained by the handbook method are averaged as explained in Section 6.3.1. The results suggest, that the assumption of altitude independence can be adhered to in the concept and preliminary design phase of aircraft, whereas in detailed design an exhaustive investigation of the temperature field is required.

Table 6.3: Temperature analysis: ratio of $\frac{\Delta T}{T_\infty}$

| $Ma/-$ | $\Delta T/T_\infty$ | | | %-difference |
|---|---|---|---|---|
| | Equation 6.4 | $alt = 10km$ | $alt = 20km$ | $alt = 10km$ to $H = 20km$ |
| 1,4 | 0,3432 | 0,351918 | 0,350034 | 0,536 |
| 1,6 | 0,448 | 0,459837 | 0,457030 | 0,611 |
| 1,8 | 0,567 | 0,581977 | 0,565862 | 2,769 |
| 2,0 | 0,700 | 0,718284 | 0,707469 | 1,506 |
| 2,2 | 0,847 | 0,837829 | 0,837250 | 0,069 |
| 2,5 | 1,094 | 1,081871 | 1,080030 | 0,170 |
| 2,75 | 1,323 | 1,405457 | 1,377302 | 2,003 |
| 3,0 | 1,575 | 1,559665 | 1,551180 | 0,544 |

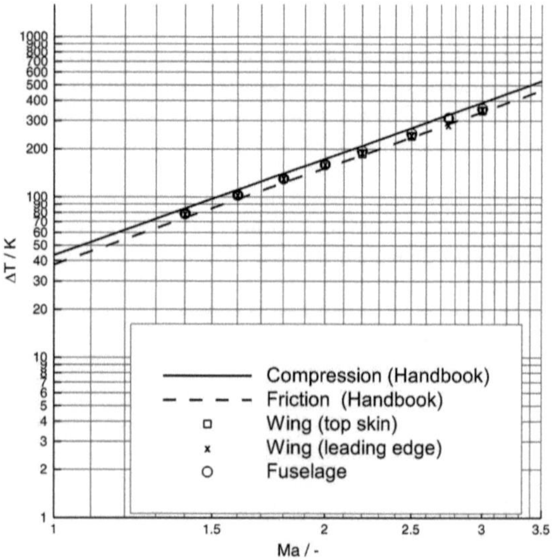


**Figure 6.5:** Temperature analysis: comparison of handbook methods and CFD-data for different Mach numbers at constant altitude ($10km$) and constant angle of attack ($0deg$).

The influence of the angle of attack is investigated by computing solutions between $\alpha$ =-2.0$deg$ and $\alpha = 6.0deg$ at constant Mach numbers of $Ma = 1.6$ and $Ma = 2.2$. Figure 6.6 shows the result for $Ma = 2.2$, $alt = 10km$ and $\alpha = 6.0deg$. One can observe that the temperature at the bottom surface increases with an increasing angle of attack, whereas it decreases at the top surface. A movement of the stagnation point cannot be observed, and consequently the qualitative temperature distribution does not change. The influence of the change of the angle of attack on the empennage is negligible. A Mach number of $Ma = 1.6$ results in a temperature increase of about $\Delta T = 0.7K$. At a Mach number of $Ma = 2.2$, the value is $\Delta T = 1.5K$. The change at the upper surface has the same magnitude but is negative, hence a temperature drop.

### 6.3.3 Influence of the Results on Materials Selection

As outlined in the introduction to this chapter, the primary goal of the investigation on temperature is to answer at which Mach number temperature effects due to speed have to be considered in conceptual and preliminary aircraft design. Consequently, the result of the described analysis is linked to material data. In the materials selection process not only data on strength and endurance is required, but information on fatigue, fracture toughness, creep data and corrosion is desired, where this listing is not complete. In the case presented here, isotropic material is solely judged on ultimate tensile strength and yield strength under tension, hence static strength. MIL-HDBK-5J (2003) gives concise data on these values and therefore uncertainties or misunderstandings resulting from data of multiple sources are avoided. However, the influence of temperature on material data will be investigated in future approaches to this subject. In Figure 6.7 the temperature due to friction and compression is plotted as calculated from handbook methods. Additionally, the allowed temperature for different materials is shown in the plot and allows to link a Mach number limit to each material. The filled symbols represent the ultimate tensile strength of the material, whereas the others show the yield strength. This comes in handy when selecting material in the conceptual and preliminary design phase of supersonic aircraft, since the engineer is given a rough estimate of what is possible. The data has been derived from the charts A.6, A.7 and A.8 in the Appendix. Several assumptions have to be made to obtain the material data:

1. The approach is only valid for large surfaces
2. Local extrema are neglected and have to be analysed separately
3. The thermal behaviour of material combinations is not regarded
4. It is assumed that a reduction of 20% of the strength values is acceptable [1]

[1]It is quite common in conceptual and preliminary design that strength values are reduced by up to 70%

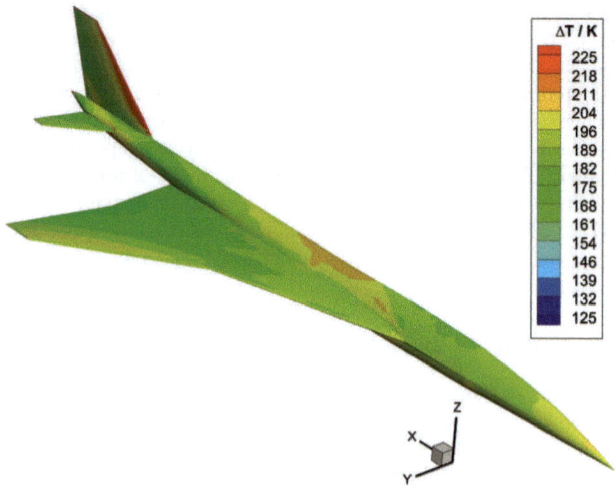


(a) top surface

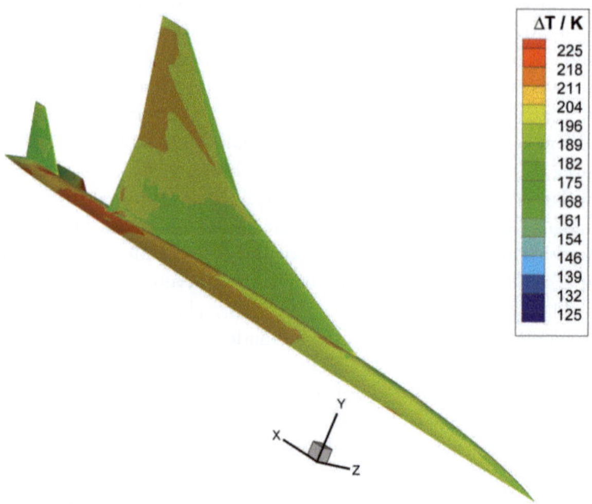


(b) bottom surface

**Figure 6.6:** Temperature analysis: temperature distribution at $Ma = 2.2$, $alt = 10km$ and $\alpha = 6.0deg$

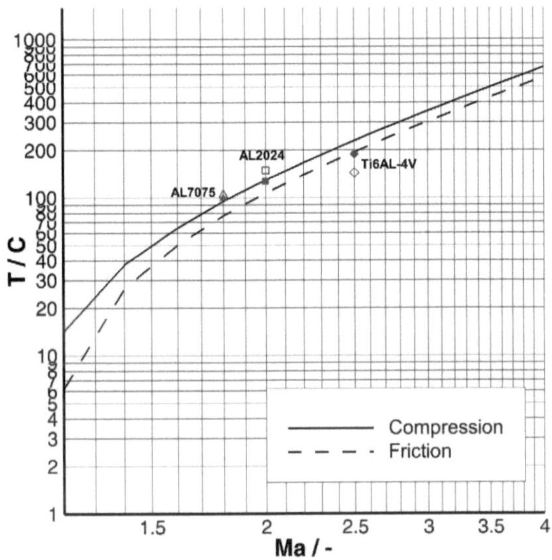


**Figure 6.7:** Temperature analysis: material surface temperature limits in Celsius and resulting flight Mach numbers. Handbook values for friction and compression (altitude = $10km$, angle of attack = $0deg$).

By way of example, the data for the commonly used aluminium alloy AL2014[2] is described here. Figure A.6 shows the reduction of tensile strength and yield strength due to the influence of temperature. It is assumed that the aircraft herein presented is subject to more than 10000 hours of thermal loading. A reduction of ultimate tensile strength of 20% is reached at about $260°F = 112°C$, whereas a temperature of about $300°F = 134.6°C$ is needed to reduce the yield strength by this percentage. The resultant Mach number is Ma=2.0. Consequently, one can state that AL2024 can be used below Mach numbers of Ma=2.0. A detailed investigation is necessary for higher Mach numbers and for applications that do not fulfil the above-mentioned assumptions, e.g., the investigation of creep effects. More detailed information and an approach for composite materials is given by Bekemeyer (2014).

[2]The data used here, is for AL2024-T3, T351 and T4 as given by MIL-HDBK-5J (2003)

# 7 Application of the Preliminary Design Process

Now that all methods are set in place, they are used to design a Supersonic Business Jet (SSBJ). Therefore, firstly some peculiarities of the business jet market and an evaluation methodology are described in short in Section 7.1. A set of requirements is derived and presented in Section 7.2, which is followed by the description of the basic design and the corresponding results in Section 7.3. Section 7.4 provides results of parameter studies, especially the variation of the design Mach number and the design range, and is followed by conclusions. Finally, the performance of the resultant aircraft on missions differing from the design mission is analysed. The focus lies on combinations of subsonic and supersonic Mach numbers in cruise flight since this is of interest, as will be outlined in Section 7.2.

## 7.1 The Market and an Evaluation Methodology

The market for business jets is driven by individuals and organizations that value their flexibility and do not want to depend on fixed airline schedules. Business jets are often considered to be flying offices, providing safety, privacy and a good working environment. The predominant factor to use a business jet is that a substantial amount of time can be saved, which is appealing to high net-worth individuals or when running on a tight schedule. Finally, the use of a business jet is often linked to prestige, imposing a lasting impression on business partners or guests. Let all the above factors be grouped as *soft factors*. When a time benefit is considered to be predominant, then supersonic travel might be the next step for some individuals and organizations that currently travel subsonic. Liebhardt & Lütjens (2011) give detailed insight into the matter and modes of operation.

Given that the hypothesis of an existing market for supersonic business jets is true, the question arises on how an aircraft design is evaluated as a good or bad design. From a financial perspective, the predominant factors are most certainly ownership and fuel consumption. These can be estimated in standard cost models such as models for Direct Operating Costs (DOC). While ownership can be quantified by the cost of the aircraft[1], fuel consumption is influenced by operating models. Operating models are well described by Liebhardt & Lütjens (2011) and are not further discussed since this is out of scope of this thesis. Compared to a rather homogeneous airliner market with classic network carriers and low-cost carriers the market for business jets is rather heterogeneous, which makes the evaluation of benefits more difficult. While airlines use, e.g., revenue seat kilometres, business jet owners and

[1] In the case of a SSBJ the aircraft cost will certainly be driven by low production rates and high development costs.

operators might identify additional sources that influence revenue. Time saving and comfort are among such sources and are difficult to measure as their value depends heavily on organisational structures and subjective feelings (soft factors). Developing such models is also not the task of this work. An impressive text on supersonic airliner operations is provided by Davies (1998).
But how does the above information relate to the forthcoming considerations? – Knowing that a standard DOC model might not be the best way to evaluate a (supersonic) business jet, it still provides a good approximation to capture the effects which are related to supersonic aircraft as will been shown in the upcoming sections. The use of this model is acceptable as long as the results are assessed relatively to each other. However, the above considerations have to be kept in mind in the course of the following sections. In the end, *soft factors* might be more relevant than cost.
The economic impact is measured by Direct Operating Costs (DOC) in Euro per seat kilometre (SKM) calculated for the design mission as outlined by Heinze (1994) and Werner-Westphal et al. (2008a).

$$DOC = \frac{\sum_{i=1}^{n} C_i}{\sum RSKM} \tag{7.1}$$

Equation 7.1 relates the total costs of the aircraft over its lifetime to the revenue seat kilometres ($RSKM$) flown during its life cycle. Costs that depend on the aircraft's utilization, i.e. fuel costs, crew costs, costs for maintenance, repair and overhaul (MRO) as well as fees and charges are calculated for a representative mission, which in this study equals the design mission. This basic equation will be used in the subsequent sections.

## 7.2 Requirements

### 7.2.1 Remarks on the Supersonic Overland Flight Ban

In many countries, legislation restricts supersonic flight over land. In the United States, civil supersonic flight is banned entirely (FAR 91, Section 817). Other countries do not allow the boom to reach the ground. In this case, supersonic flight up to a "cutoff Mach number" is allowed (For Germany see LuftVO Section 11a). This Mach number is usually around Ma=1.15, depending on atmospheric conditions. Consequently, this legislation has to be taken into account when considering the development of supersonic aircraft. Primarily, the laws have an impact on flight routing and are therefore a key in market analysis, as Liebhardt et al. (2013) state.

### 7.2.2 Cruise Mach Number

A SSBJ's primary purpose is to fly at high speeds. In the case presented here the aircraft must be able to cruise between Mach 1.2 and 1.7. The lower limit is derived by the desire not to fly in the transonic regime, hence to avoid the area of high aerodynamic drag. The limit of Mach 1.7 as upper value is not driven by operational considerations but rather by technical feasibility issues regarding the propulsion system. The purpose of the propulsion system is to accelerate the aircraft to cruise speed. Engines with high jet velocities are needed to achieve high cruise Mach numbers and thus low bypass ratio engines have to be selected. The drawback of this engine type is its high noise at take-off. To satisfy regulations on noise, engines with high bypass ratios are needed, hence, engines with large cross sections that consequently produce higher aerodynamic drag. At this point, the contradiction becomes clear. Furthermore, at Mach numbers higher than 1.7 a variable engine inlet is needed to avoid huge pressure losses that would result in poor engine efficiency. Such inlets are complex and require a high degree of maintenance. For this reason, they are not an option for a SSBJ. Additionally, results presented in Chapter 6 suggest that Mach numbers below Ma=2.0 are advisable to avoid an undesired heating of the structure. As a result from the above considerations and based on comparison with other SSBJ projects (cf. Section 1.2) Ma=1.6 is selected as design Mach number for the aircraft presented here.

### 7.2.3 Range, Payload and Cabin

#### Range

Liebhardt et al. (2011) analysed the market potential for supersonic transportation. The authors therefore, evaluated the worldwide premium ticket sales figures. The basic assumption of this study is that a certain percentage of premium passengers would be willing to change to supersonic service and hence pay the higher airfare. Design ranges of 3585 nm (6640 km), 4265 nm (7900 km) and 5065 nm (9380 km) for conventional supersonic business jets are suggested, given that the supersonic overland flight ban (cf. Section 7.2.1) will not cease to exist. If these ranges were selected, approximately 54%, 60% and 65% of the potential supersonic market could be covered by non-stop connections. The notion of a *conventional supersonic aircraft* in this context means an aircraft that has to comply with the current supersonic overland flight ban, hence flies only over water at supersonic speeds, and does not encompass sonic boom mitigation technologies. In this case, so called boom-restricted routing would be necessary (see Liebhardt et al. (2013) for more). Should a low-boom design be developed the figures would change according to Liebhardt et al. to 3435 nm (6360 km) and 5350 nm (9720 km), respectively, with a market coverage of approximately 45% and 80%.

Based on this analysis, and existing supersonic business jet projects (cf. Section 1.2), a few conclusions regarding the range requirement can be drawn. Firstly, the aircraft must be capable of serving at least transatlantic routes. Secondly, the market potential increases with an increasing range, since routes in the Pacific Rim would become accessible. However, routes with one stop are acceptable when still justified by time savings. This limits the design range to a maximum. As outlined in Section 7.2.2, a cruise Mach number below Ma=1.2 and therefore flying below the "cutoff Mach number" is not sensible. Hence, overland routing in Europe or Asia is not an option. Table 7.1 shows a selection of city pairs and their corresponding great circle distances that are interesting from a business perspective. They are illustrated in Figure 7.1. The city pair Los Angeles-Tokyo has been added as being representative for the Pacific Rim. It can be seen clearly for which routes boom-restricted routing or partial subsonic cruise would be necessary. At the current stage of the development, PrADO does not provide the capability to asses the sonic boom. Therefore, no boom mitigation measures are integrated into the proposed design. Consequently, the aircraft will be considered as conventional SSBJ. Reflecting on what has been said above, a design range of 4000 nm (7408 km) as chosen for the HISAC aircraft seems to be reasonable and is therefore selected for the following studies.

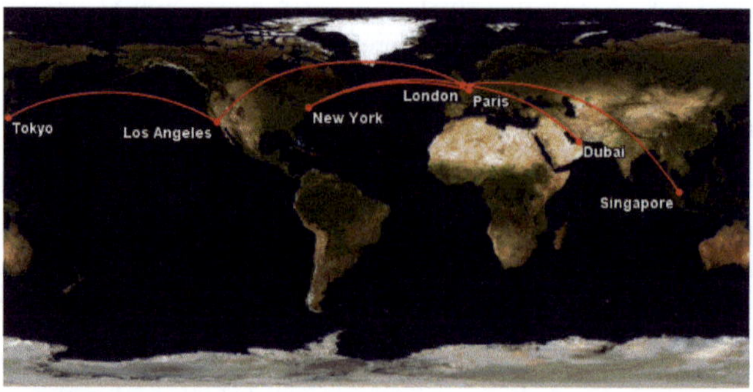


**Figure 7.1:** Selected city pairs and great circle routes (gcmap.com)

**Table 7.1:** Top five city pairs and great circle distances based on premium ticket revenue, Liebhardt et al. (2013)

| *Origin* | *Destination* | *GC D. /km* |
|---|---|---|
| London | New York | 5,555 |
| London | Singapore | 10,886 |
| London | Dubai | 5,504 |
| Paris | New York | 5,848 |
| London | Los Angeles | 8,781 |
| | | |
| Los Angeles | Tokyo | 8,773 |

### Payload and Cabin

After speed and range, the amount of payload needs to be determined to complement the basic requirements. Typical larger business jets do provide space for eight to nineteen passengers[2], whereas the low number represents a more comfortable arrangement than the higher one. Based on this assessment, the design point ot the aircraft presented herein is chosen to be eight passengers. The maximum payload is derived from the upper limit and chosen to be nineteen passengers. Passenger weight is set to 77 kg per passenger and an average of 14 kg of luggage per passenger is assumed. Consequently, the payload at the design point equals 728 kg and maximum payload equals 1728 kg. One attribute of a business jet is a cabin that is often more comfortable than business class or first class arrangements of airlines. The selected layout is depicted in Figure 7.2.

### 7.2.4 Additional Requirements and Requirements Summary

The discussed requirements are summarized and complemented by additional data in this section in Table 7.2. The acronym ETOPS refers to the *Extended-range Twin-engine Operational Performance Standards* issued by the International Civil Aviation Organisation (ICAO) . The standard applies to twin-engined aircraft and the time that they are allowed to fly with only one engine running.

**Table 7.2:** Requirement summary

| *Parameter* | *Value* | *Unit* | *Remarks* |
|---|---|---|---|
| Number of Passengers (design point) | 8 | - | - |
| Passenger mass (incl. luggage) | 91 | kg | - |
| Maximum Payload | 1728 | kg | - |
| Range at design point | 4000 (7408) | nm (km) | - |
| Cruise Mach number | 1.6 | - | - |
| $V_{AT}$, speed at threshold | $< 140$ | kts | cf. Section C.2 |
| Max. allowed runway length | 2200 | m | serve smaller airports. |
| Max. cruise altitude | $FL600$ | - | avoid interference with normal air traffic. Same as for Concorde. |
| ETOPS | 180 | min | allow to serve North and Latin America |

[2]e.g. Bombardier Global 6000, Dassault Falcon 7X, Gulfstream G650

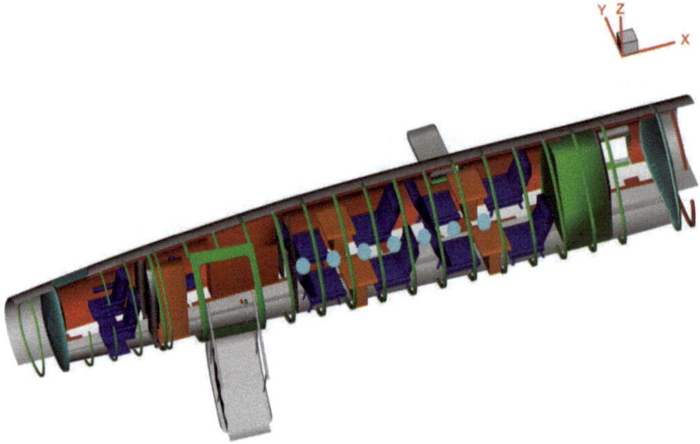


(a) Selected eight-passenger arrangement - PrADO representation

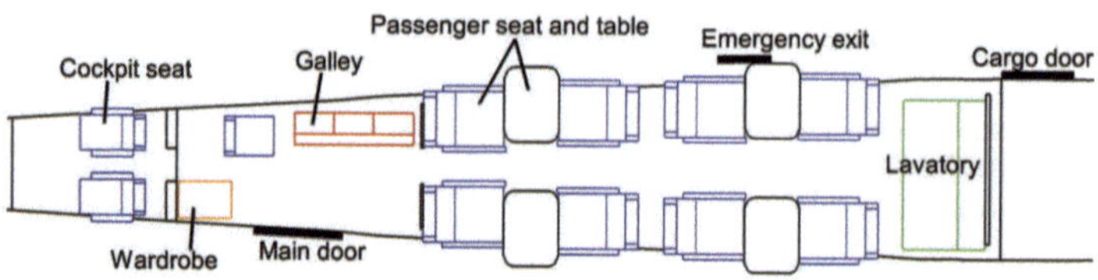


(b) Selected eight-passenger arrangement - 2D drawing

**Figure 7.2:** Selected cabin arrangement

## 7.3 Basic Aircraft

The goal of the analysis is to determine the Maximum Take-off Mass for a given payload and range, while keeping the shape of the aircraft fixed. For this purpose, the aircraft model is described and results are presented.

### 7.3.1 Model Description

The aircraft is designed according to the requirements and specifications in Section 7.2. Two major changes have been made compared to the HISAC configuration in Chapter 5.3. The surface of the canard has been increased from $6.1m^2$ to $7.0m^2$ ($\approx 14.7\%$) and the distance between the aerodynamic centre of the wing and the canard has been increased marginally. Both changes contribute to better stability. Additionally, the increased surface influences the overall lift positively[3]. The second change is the most visible one. The configuration proposed herein comprises only two engines. This change allows for a smoother rear fuselage and the possibility of increasing the fuel volume. The engine's accessibility is enhanced, since both engines are under the wing and can therefore be accessed easily. However, the two engines have to provide more total thrust than the three engines did, when assuming that OEI (one-engine-inoperative) is the sizing criterion. Hence, they are expected to be bigger and the ETOPS certification is required. The OEI case will have an impact on the sizing of the vertical tailplane. This is due to an increased yaw moment when compared to the configuration with three engines that is caused by the bigger engines. After all, the used version of PrADO does not size the area of the vertical tailplane, thus it is based on engineering judgement. Some geometrical data is given in Table 7.3. Further data is shown in Tables B.10, B.11, B.12 and B.13 of the Appendix. A 3D view of the aircraft is provided in Figure 7.3 and 2D views are to be found in Figures A.10 of the Appendix.

**Table 7.3:** Basic aircraft: geometrical data

| *Parameter* | *Value* | Unit |
|---|---|---|
| Aircraft length, $l_{AC}$ | 36.80 | $m$ |
| Wing area, $A_W$ | 150.0 | $m^2$ |
| Wing span, $s$ | 18.50 | $m$ |
| Leading edge sweep angles, $\varphi$ | 72.5, 52.0 | $deg$ |
| Dihedral angles, $\Gamma$ | 0. | $deg$ |
| Number of engines, $n_E$ | 2 | – |

The selection of materials is based on information available from the HISAC project (cf. Deremaux (2009)). The primary structure of the aircraft consists of carbon-fibre reinforced plastics (CFRP) for the cover sheets of the fuselage and the lifting

[3]In contrast to a horizontal tailplane (HTP), a canard contributes to lift.

surfaces. Aluminium alloy is used for other structural elements. Material data is given in Table B.21. The selected material and its knock-down factor, $\phi_M$, for each subcomponent are given in Table B.22. The knock-down factor is estimated based on aircraft that have previously been analysed with PrADO (see e.g. Österheld (2003); Rieke (2013)). The SSBJ comprises six fuel tanks in the inner part of the wing and two fuel tanks in the fuselage, providing a capacity of $44.7m^3$, which equals $36132.52kg$[4]. The aircraft has one main passenger door and an emergency exit. An additional cargo door is integrated on the starboard side right after the passenger cabin and before the rear pressure bulkhead. The cargo area provides a volume of $2.30m^3$ for luggage. The undercarriage is composed of a nose gear strut with two wheels and two main gear units with four wheels each. The nose gear retracts completely into the fuselage underneath the canard, whereas the main landing gear struts retract into the wing and the wheels are positioned between the forward and rear fuel tank of the fuselage.

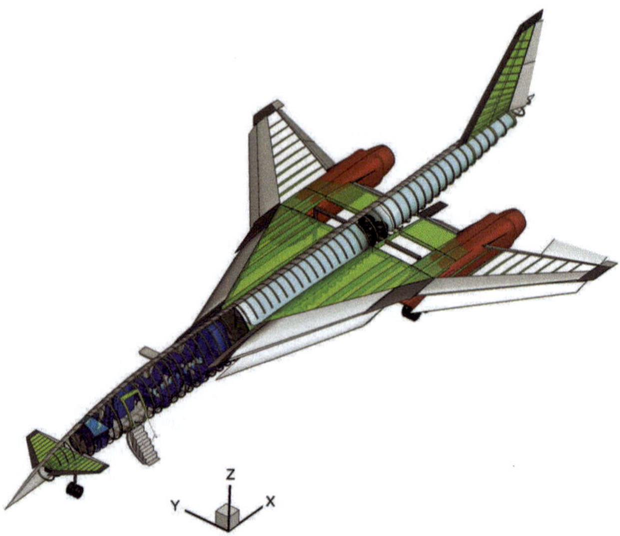


**Figure 7.3:** Basic aircraft: PrADO representation

[4]Detailed data on the used kerosene can be found in Table B.23

### 7.3.2 Results

Results of the analysis are presented in Table 7.4. The masses are justifiable for an aircraft of this type and size. Cruise altitudes for the design mission and the runway length at landing fulfil the requirements. The maximum wing loading is a reasonable value. A low wing loading increases manoeuvrability and will enable the aircraft to take-off and land at lower speeds. This goes along with the speed at threshold, which is lower than its upper limit. Figures for other aircraft are provided in Table B.20 to give an idea of this quantity. A short theoretical introduction to wing loading is given in Appendix C.3. For conventional civil aircraft the design case for the engine is most commonly engine failure at take-off, hence a OEI (one-engine-inoperative) scenario. In contrast to this it is observed that the engine of this aircraft is sized according to the cruise condition. This is due to the high speed and thrust required. The specific fuel consumption under cruise conditions, $SFC_{Cruise}$, is determined to be $0.08555\frac{kg}{Nh}$. To get a feeling for the quantity: the specific fuel consumption under cruise conditions for an Airbus A318 which has a $m_{MTOM}$ of 54823kg, hence comparable to the SSBJ described herein, is $SFC_{Cruise} = 0.05616$[5]. Additional information on the propulsion system is given in Table B.15 in the Appendix. To get a feeling for the DOC value in Table 7.4: For an Airbus A330-300, hence an aircraft capable to cross the Atlantic, the value is $DOC = 0.0289\frac{\text{EURO}}{SKM}$[6]. Even thought the SSBJ and the Airbus A330 are not comparable aircraft, neither by size nor by their target market, the values give an impression of the difference in the order of magnitude.
Figure 7.4 shows the general behaviour of the design iteration for various masses, static thrust and range. The following global parameters terminate the iteration at convergence:

maximum take-off mass, $m_{MTOM}$

the operating empty mass, $m_{OEW}$

fuel mass of the design mission, $m_{fuel,D}$

static thrust per engine, $S_{0,ENG}$

Convergence requires that the relative change of these values is less than or equal to $1 \times 10^{-4}$. This is true for all values after fifteen iterations. The ranges (Figure 7.4(c)) are merely specified for reasons of completeness, but it can be observed that they converge well. The indices 1 to 3 indicate missions with maximum payload, maximum fuel and the ferry flight, whereas index 4 is the design mission, thus the constant range.

[5] The data has been derived with PrADO based on methods introduced in Österheld (2003). The design range is 2778 km at $Ma = 0.76$.

[6] The data has been derived with PrADO based on methods introduced in Österheld (2003). The design range is 8898 km at $Ma = 0.82$ and $m_{MTOM}$ of 233983kg.

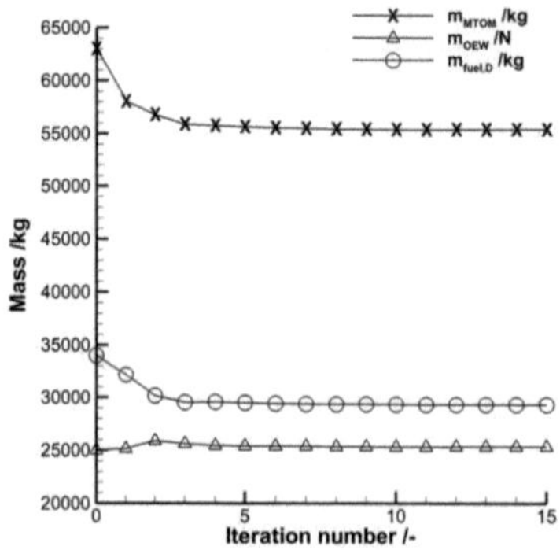


(a) Mass

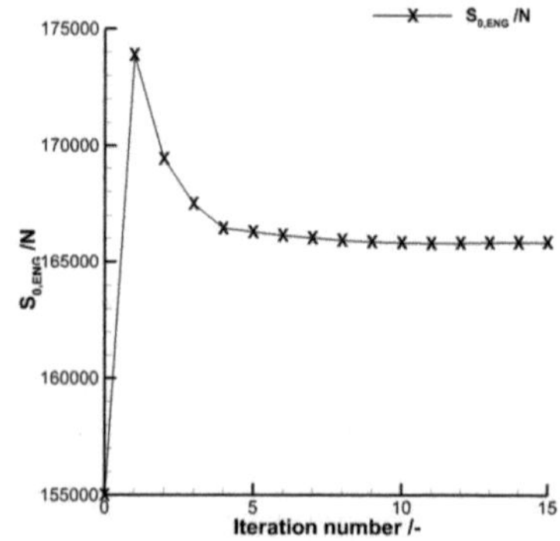


(b) Static thrust

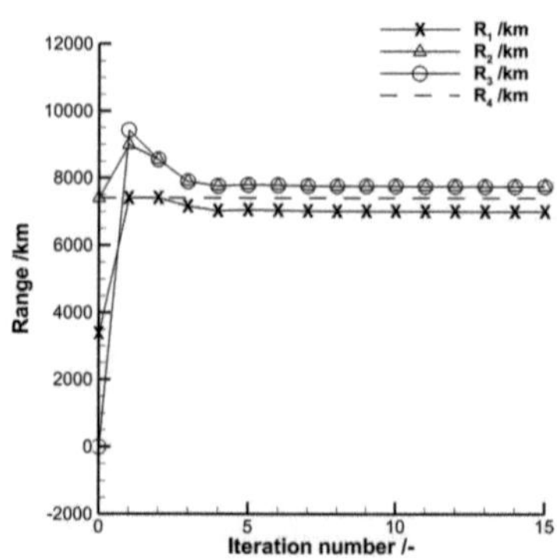


(c) Range

**Figure 7.4:** Basic aircraft: global iteration behaviour

**Table 7.4:** Basic aircraft: PrADO results

| *Parameter* | *Value* | *Unit* | *Remark* |
|---|---|---|---|
| $m_{MTOM}$ | 55412. | $kg$ | Max. take-off mass |
| $m_{OEW}$ | 25352. | $kg$ | Operational empty mass (weight) |
| $m_{fuel,max}$ | 30058. | $kg$ | Max. fuel mass |
| $W_{S,max}$ | 369.41 | $\frac{kg}{m^2}$ | Max. wing loading |
| $S_0$ | 333.92 | $kN$ | Total static thrust |
| $ALT_{begin,cruise,D}$ | 14.341 | $km$ | Altitude at the begin of cruise at the design point |
| $ALT_{end,cruise,D}$ | 17.955 | $km$ | Altitude at the end of cruise at the design point |
| $L/D_{begin,cruise,D}$ | 5.59 | $m$ | Lift-to-drag ratio at the begin of cruise at the design point |
| $V_{AT}$ | 243. , (131.2) | $\frac{km}{h}$, $(kts)$ | Speed at threshold |
| $l_{RWY,T}$ | 2485. | $m$ | Runway length at take-off |
| $l_{RWY,LD}$ | 1484. | $m$ | Runway length at landing |
| DOC | 0.44630 | $\frac{\text{EURO}}{SKM}$ | Direct operating costs |

## Aerodynamics

Aerodynamic data is presented in Figure 7.5 for untrimmed cruise conditions. For reasons of clarity, only three altitudes at a constant cruise Mach number of $Ma = 1.6$ are plotted. The selected altitudes are $10km$, $15km$ and $18km$, whereas the two last mentioned values are chosen based on the values given in Table 7.4. Total drag as function of the angle of attack is represented in plot 7.5(a), and showing the expected behaviour. The altitude dependency can be better observed in the zoom in detail of Figure 7.5(a). Lift as function of the angle of attack is presented in Figure 7.5(b). Lift is not as altitude-dependend as drag, and its dependency is barely visible in the plot. The linear behaviour is commonly expressed as in equation 7.2. Here $C_{lift,\alpha} = \frac{dC_{lift}}{d\alpha}$ denotes the lift curve slope, $\alpha_0$ is the angle for zero lift and $C_{lift,0}$ denotes zero lift, hence the lift produced when the angle of attack is zero.

$$C_{lift} = C_{lift,\alpha} \cdot (\alpha - \alpha_0) = C_{lift,\alpha} \cdot \alpha + C_{lift,0} \tag{7.2}$$

$$C_{drag} = C_{drag,0} + k \cdot C_{lift}^2 \tag{7.3}$$

The aerodynamic drag polar is given in Figure 7.5(c). This plot follows the assumption of a quadratic polar as expressed in equation 7.3, where $C_{drag,0}$ denotes the zero drag coefficient and and the quadratic term is the lift-induced drag.
The lift/drag polar is given in Figure 7.5(d). This plot allows determining the maximum lift-to-drag ratio, which is in this case close to the value 6.5. It can be seen that the computation of more data points would be desirable to achieve a better representation of the curve. For reasons of completeness, the pitching moment coefficient is given in Figure A.11 of the Appendix. The behaviour of both graphs meets the expectations.

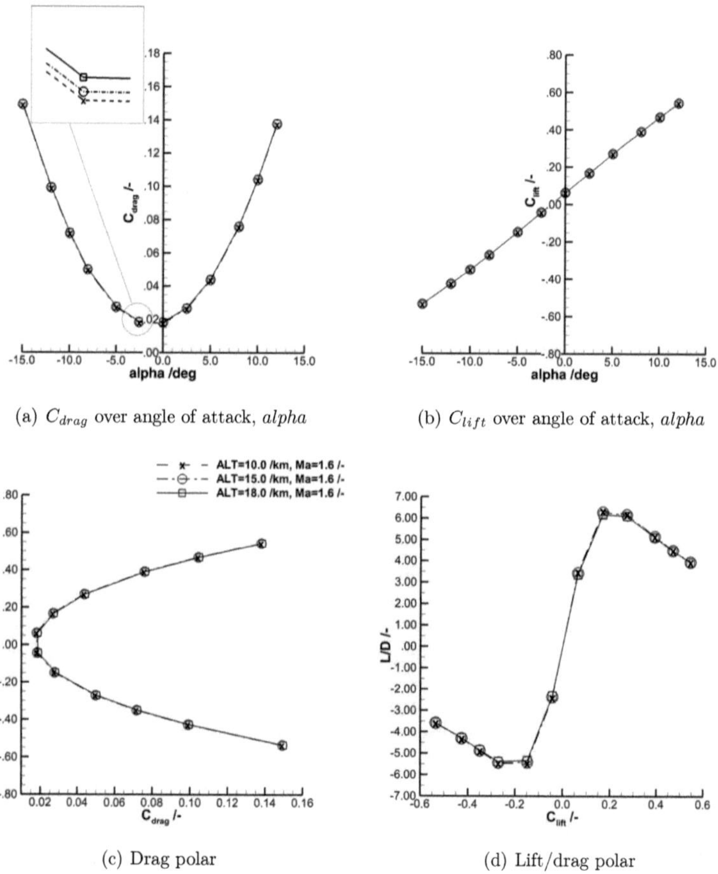


(a) $C_{drag}$ over angle of attack, *alpha*

(b) $C_{lift}$ over angle of attack, *alpha*

(c) Drag polar

(d) Lift/drag polar

**Figure 7.5:** Basic aircraft: aerodynamic data

Table B.17 shows that the lift-to-drag ratio at the beginning of cruise is computed to be 5.6. The reduction compared to the value mentioned above is due to trim effects, notably trim drag. Figure 7.6 shows the drag polar for various deflections of the trim device. It is visible that a deflection of the canard moves the polar towards higher drag coefficients and lower lift coefficients. This is due to the trim drag and results in lower lift-to-drag ratios than for a clean configuration.

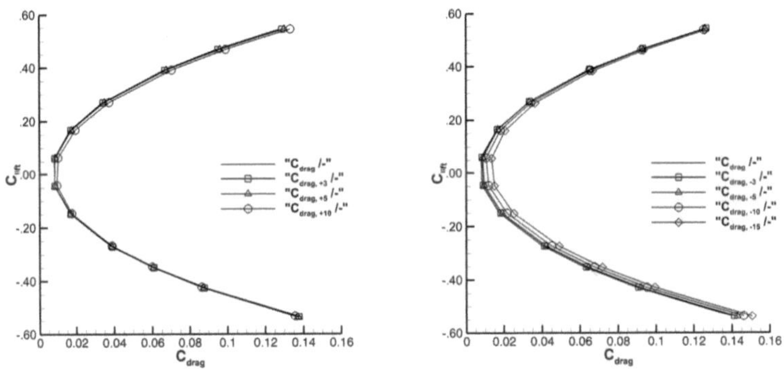


(a) Drag polar for various positive trim deflections

(b) Drag polar for various negative trim deflections

**Figure 7.6:** Basic aircraft: influence of trim deflections on the the drag polar at $Ma = 1.6$ and constant altitude

### Operating Empty Mass

A mass break down of the operating empty mass is given in Table B.14 of the Appendix. The mass of the propulsion system is derived from the thermodynamic analysis (cf. Section 3.3). All other values, except for the primary structure (parameter numbers 1-4 in the table), are calculated using PrADO standard methods. These methods are well established in PrADO and most frequently based on suggestions by Heinze, Howe, Raymer and Torenbeek. The mass of the primary structure is based on finite-element analysis as explained in Sections 3.2 and 4.2. The load cases that are taken into account for the design are presented in Table B.16. It has to be noted, that the values presented are valid for the final design iteration only. Data for previous steps is calculated as needed and differs from the values in the table, since aircraft mass and performance data change from iteration step to iteration step until convergence is achieved. The convergence behaviour of the structural sizing model is presented in Figure 7.7 where it can be seen that both convergence criteria are satisfied after eight iterations. Convergence of the maxi-

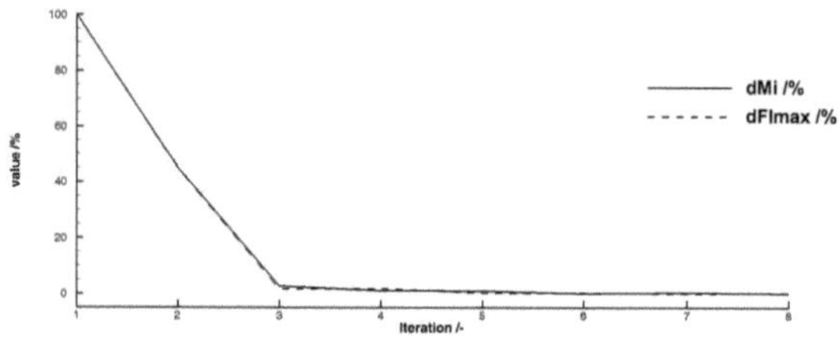


**Figure 7.7:** Basic aircraft: convergence of structural sizing for parameters $dFI_{max}$ and $dMi$

mum failure index is reached after five iterations, whereas convergence of mass is reached at iteration eight. Note that the ordinate starts at a value lower than 0.0 to enhance the visibility of the data. Default settings (cf. Table B.2) were applied for this aircraft, since they proved to give fine results during verification (cf. Section 5.2).

The load cases are only *nearly* symmetric. *Nearly* in this context requires some explanation: The load that is exerted on the aircraft is not absolutely symmetric. Symmetric loading is only achieved for aerodynamic loads, fuel masses and landing gear loads. All aerodynamic load cases are symmetric to the xz-plane. Inertia loads are not always symmetric. Consider the galley, for example. It is in this case only installed on the starboard side of the cabin, whereas the passenger door is opposite. Consequently, the load resulting from these masses differs and is not symmetric. This can happen for any other installation like systems, cabin interior, and the like and must not only be true for the fuselage. Figure 7.8 shows the resultant skin thickness from different perspectives. As it would be expected, the distributions are symmetric along the xz-plane. However, some deviations can be observed on the bottom skin in Figure 7.8(b). These deviations cannot be explained by the asymmetric inertia loads. Since the implementation has been verified in Section 5.2, it is most likely that the deviation is caused by either a local modelling issue or is an outlier. A detailed look revealed, that the aerodynamic forces transferred onto the model cause the difference on the bottom skin. Other aircraft available in PrADO have been examined based on this discovery. The thickness distributions of these aircraft show to be symmetric, indicating that the issue is specific to the SSBJ aircraft. To illustrate this, the skin thickness distribution for the SFB880 aircraft (Weiss & Heinze (2013)) is given in Figure A.12 as an example without further discussion. However, the impact of this issue on the structural mass is negligible, since the mass difference between the starboard part of the wing and the portside part of the wing is less than $8.0kg$. Remember, the goal of structural sizing in preliminary aircraft design is to obtain global aircraft masses and these masses seem reasonable.

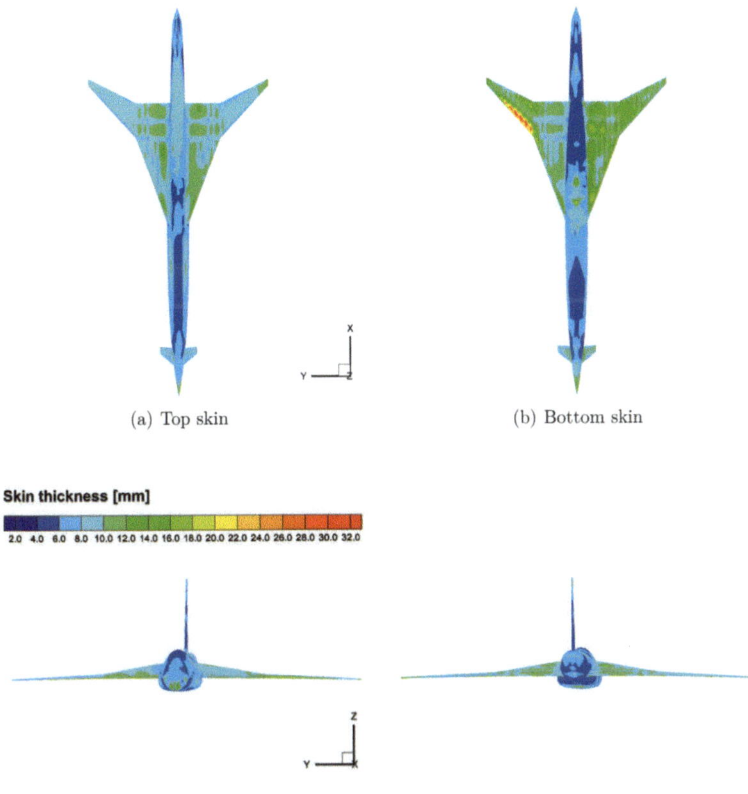


(a) Top skin

(b) Bottom skin

(c) Front view

(d) Rear view

**Figure 7.8:** Basic aircraft: skin thickness obtained with SSM

### Mission Simulation and Mass-Range Chart

Masses and ranges for the mass-range chart are the result of a flight simulation. The mission outline is depicted in Figure 7.9. Altitude, Figure 7.9(a), and Mach number, Figure 7.9(b), are shown over the range. It is modelled from different segments, each segment describing a flight phase. The cruise segment is modelled as a cruise-climb flight. Segments for taxi, take-off, and climb are modelled, as well as an acceleration phase at constant altitude. The final phase of the flight consists of descent, approach and landing. Additionally, the flights in a holding pattern and to an alternate destination are simulated to determine the reserve fuel mass. The range of the design mission is a given quantity and the lenghts' of the mission segments are adjusted to match it. Major quantities resulting from the simulation of the design mission are the maximum take-off mass $m_{MTOM}$ and the fuel mass for the mission. The same mission profile is used to simulate the flight with maximum payload, maximum fuel and the ferry flight. In contrast to the design mission, not only masses are a result of the analysis, but ranges are calculated and complement the data. Eventually, the results allow to drawing a mass-range chart.
It is depicted in Figure 7.10. The chart shows the expected behaviour up to the design point. The range for the flight with maximum payload is $7000km$. The chart differs from what one would expect for a conventional aircraft at the right side of the design point. Here, it can be seen that the ranges for a flight with maximum fuel and for a ferry flight are close together. In fact, they are the same and no payload can be carried when flying with maximum fuel. The reason for this behaviour can be explained by analysing the global aircraft masses. The take-off mass ($m_{TOM}$) is the sum of the operating empty mass, payload and fuel mass. This is shown in equation 7.4. The maximum fuel volume is usually defined as the maximum volume available for fuel. For this aircraft the value is $V_{fuel,max} = 44.673m^3$, which equals to a mass of $36131.52kg$ based on the density defined for kerosene in Table B.23. For a flight with maximum fuel one would now calculate the payload, $m_{payload}$, which can be carried by rearranging equation 7.4 and substituting $m_{TOM}$ by $m_{MTOM}$.

$$m_{TOM} = m_{OEW} + m_{fuel} + m_{payload} \tag{7.4}$$

The used values are:

$$m_{MTOM} = 55412.kg$$

$$m_{OEW} = 25352.kg$$

$$m_{fuel} = 36131.52kg$$

The result is a payload of $m_{payload} = -6071.kg$, which is obviously not reasonable, thus $m_{payload}$ is set to zero. Applying equation 7.4 again but this time rearranging for $m_{fuel}$ one obtains $m_{fuel} = 30058kg$ which is exactly the result shown in Table 7.4. This result can be summarized as follows: In contrast to standard passenger

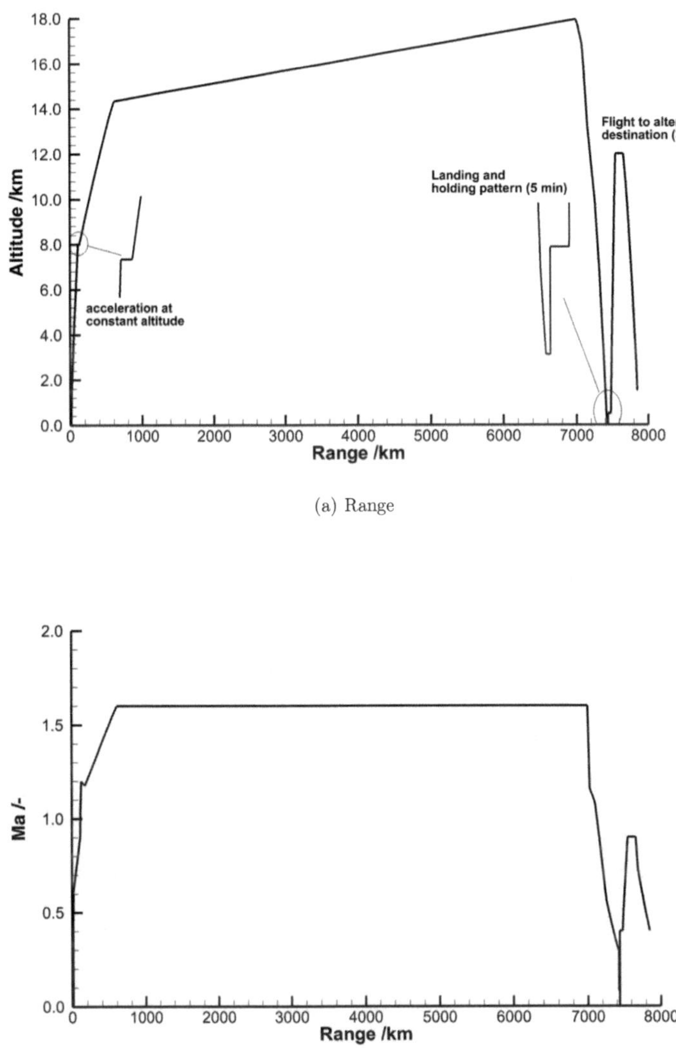


(a) Range

(b) Mach number

**Figure 7.9:** Basic aircraft: mission profile

aircraft the maximum fuel mass is limited by the maximum take-off mass, where usually the maximum available fuel volume would be the limit. The aircraft does only use 83.19% of the available fuel volume. Table B.17 shows data for all missions.

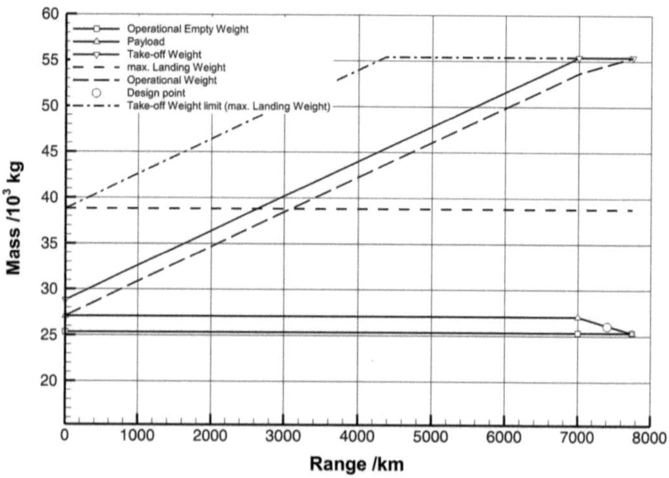


**Figure 7.10:** Basic aircraft: mass-range chart

## 7.4 Parameter Studies

Now that a basic aircraft design is available, it seems reasonable to vary parameters to show the design's sensitivity. As outlined in the requirement Section (7.2), the cruise Mach number and the design range seem to be critical for the commercial success of a SSBJ. Therefore, the Mach number and design range are varied in Section 7.4.1. Common to all variations is that the aerodynamic shape is fix, thus computational time is saved, since the aerodynamic performance map of the basic aircraft can be used. The question to answer is: is there a better combination of $Ma$ and range for the given aircraft's shape that is reasonable, and what are the mission characteristics? In a second step (Section 7.4.2) one aircraft is selected from the solution space and used for performance studies on missions different than the design mission to examine the aircraft's sensitivity on such scenarios. At this point remember: The aim of this work is not to find an optimal design, but rather to introduce methods and methodologies that allow to judge supersonic aircraft concepts.

### 7.4.1 Variation of the Cruise Mach Number and Design Range

For this parameter study, ranges between $7250km$ and $9000km$ are examined. Additionally, the Mach number is varied from 1.2 to 1.8. The economic impact is measured by Direct Operating Costs (DOC) in Euro per seat kilometre (SKM) calculated for the design mission, as outlined in section 7.1. In contrast to the basic design, it seemed reasonable to increase the payload at the design point from eight to ten passengers, thus influencing the Direct Operating Costs positively[7]. All combinations that have been analysed are marked with an X in Table 7.5. The analysis for $Ma = 1.4$, $Ma = 1.5$ and $Ma = 1.6$ and a range of $9000km$ as well as for $Ma = 1.2$ and $Ma = 1.8$ at a range of $8750km$ has shown that the required fuel mass reaches or exceeds the maximum fuel limit ($m_{fuel,limit} = 36131.52kg$, cf. Section 7.3.2) for the aircraft. Figure A.14 indicates that it is most likely that the analysis for $Ma = 1.2$ and $Ma = 1.8$ for this range would reach the fuel limit as well. The analysis of $Ma = 1.2$ and $Ma = 1.8$ is therefore not conducted at the range of $9000km$. Various data is given in Table B.18 and selected results are

**Table 7.5:** Parameter Variation: combinations of range and Mach number

| | Ma /- | | | | |
|---|---|---|---|---|---|
| R /km | 1.2 | 1.4 | 1.5 | 1.6 | 1.8 |
| 7250 | X | X | X | X | X |
| 7500 | X | X | X | X | X |
| 7750 | X | X | X | X | X |
| 8000 | X | X | X | X | X |
| 8250 | X | X | X | X | X |
| 8500 | X | X | X | X | X |
| 9000 | - | X | X | X | - |

depicted in Figures A.13 to A.18. Figures A.13, A.14 and A.15 show the dependence of $m_{MTOM}$, $m_{fuel,D}$ and $m_{OEW}$. All three masses increase with increasing range and design Mach number. This behaviour is expected and common with conventional aircraft. All three masses show an area of minima along the range close to $Ma = 1.4$ and $Ma = 1.5$, which is most obvious for $m_{fuel,D}$ and $m_{MTOM}$. The operating empty mass shows more sensitivity to Mach number variations than to a change in range, whereas the Maximum take-off mass shows more sensitivity to changes of the design range than to a change of the Mach number. Predominant for static thrust $S_0$ and hence the size of the engine is the choice of the Mach number. This is shown in Figure A.16. This behaviour goes along with the observation made in Section 7.3.2 that the size of the engine is driven by cruise conditions and not, as for conventional aircraft, by take-off with engine failure. The number of global design iterations for each parameter combination is shown in Figure A.18. Some

[7] In the following text it will be shown that a reduction of $\approx 25\%$ is achieved.

solutions were derived based on an already existing one (restart) and others were started from scratch. The main statement of this figure is, that two solutions show a value of 50 iterations, which equals to the maximum number of iterations, hence no convergence is achieved. Consequently, no feasible design could be derived under the given circumstances. Figure A.17 shows that the Direct Operating Costs reach a minimum along with Mach numbers of 1.4 and 1.5 for various ranges. Consider the following example to understand this behaviour:
The cost estimation is based on a lifetime of 14 years. For a $7750km$ design mission at $Ma = 1.4$ this equals to a flight time of 54086.739 hours[8]. Component costs are directly related to component masses. The percentage of the cost components is given in Table 7.6 for selected missions, where $R_4$ denotes the design range.

**Table 7.6:** Parameter Variation: cost breakdown for selected design missions

| *Parameter* | *Unit* | *Value* | | |
|---|---|---|---|---|
| $R_4$ | $/km$ | 7750 | 8000 | 8500 |
| $Ma$ | $/-$ | 1.4 | 1.5 | 1.8 |
| Aircraft costs | /% | 18.200 | 17.699 | 16.266 |
| Fuel costs | /% | 58.709 | 60.476 | 65.287 |
| Crew costs | /% | 10.675 | 10.025 | 8.286 |
| MRO costs | /% | 10.331 | 9.701 | 8.018 |
| Fees and charges | /% | 2.086 | 2.099 | 2.143 |

This table is characteristic for all analyses that have been conducted within this study. It can be observed that fuel is the major cost factor followed by aircraft costs. Heinze (2007) specifies a fraction of 28.4% for fuel costs of a standard $A320$ passenger aircraft. An analysis of the already quoted $A318$ aircraft with PrADO gives a fraction of 14.6% for fuel costs. The evidence is obvious: In contrast to conventional aircraft, fuel costs, $C_{fuel}$, are the major contributor to overall costs for a SSBJ. To give a feeling for the quantity: One load of fuel, $m_{fuel,max}$, for the basic aircraft from Table 7.4 would roughly cost 16500 EURO[9]. Since fuel costs have been identified as the cost driver, the influence of the Mach number on it shall be illustrated in a simple exercise. Therefore, fuel costs are plotted in Figure 7.11 as a function of the Mach number at constant range. Each Mach number range combination has been analysed as explained above, hence the data is derived from similar aircraft. It can be observed from the figure that a change of the design range has less impact on $C_{fuel}$ than the variation of the Mach number, which seems to be sensible.
The sensitivity of the numerator in equation 7.1 has been linked to fuel costs. The sensitivity of the denominator, the revenue seat kilometres (equation 7.5), shall be explored now. $RSKM$ is calculated according to equation 7.5. Here, $N_{PAX}$

[8]This equals roughly a utilization of 42% over the lifetime.
[9]Data from Table B.23.

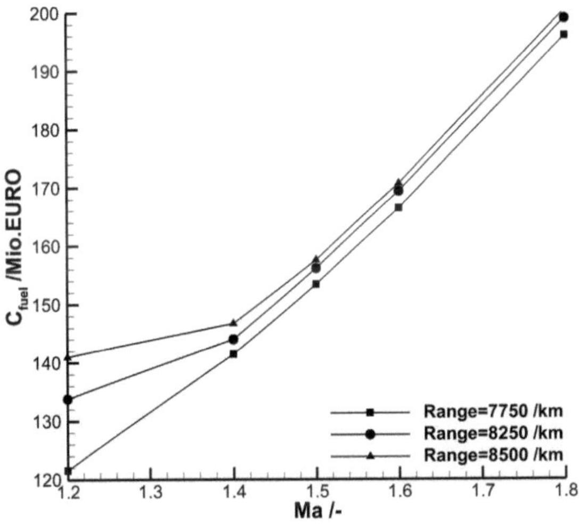


**Figure 7.11:** Parameter Variation: mach number dependence of fuel costs, $C_{fuel}$

is the number of passengers at a representative mission and $R_F$ the range of that mission. $N_F$ denotes the number of flights during the lifetime of the aircraft on the representative mission. For the analysis conducted here, the design mission and the representative mission are equal. If not done before, one could argue at this point that it is unlikely that an aircraft is used only at its design mission, and many other mission profiles must be examined to set up a valid business case. This is true. However, manufacturers and potential customers have their own DOC models, which is often proprietary information, and will apply them before undertaking an investment into an aircraft. Hence, the use of a conventional DOC model is justified at this stage of the design process to capture effects of parameter changes (cf. Section 7.1).

$$RSKM = N_{PAX} \cdot R_F \cdot N_F \tag{7.5}$$

Back to equation 7.5. It becomes obvious without further comments that a higher range and a higher number of flights during the lifetime of the aircraft[10] would influence the revenue seat kilometres positively. A change in range is easy to imagine, but the parameter $N_F$ needs some clarification. A definition of $N_F$ is given in equation 7.6

$$N_F = \frac{u_F \cdot y^* \cdot V_c}{R_F} \tag{7.6}$$

In this equation, $u_F$ denotes the annual utilization of the aircraft and $y^*$ the number of service years of the aircraft. How these quantities are derived is explained, for example, in Werner-Westphal et al. (2008a) and not further detailed here. The essence of the equations 7.5 and 7.6 is that higher speed influences the revenue seat kilometres positively and yet at a first glance $R_F$ cancels out when merging both equations into a single one[11]. $RSKM$ is plotted as a function of the Mach number at constant range in Figure 7.12. Here it becomes obvious that some Mach-range combinations are more suitable for the present aircraft geometry than others when considering revenue seat kilometres. At the bottom line: High speed seems to be favourable in terms of revenue seat kilometres but not in terms of costs. Having this in mind, the distribution of Figure A.17 is explainable. A combination of low costs and high $RSKM$ influence the DOC value positively.

The sensitivity of the payload increase on DOC shall be outlined here in brief. The design point of the basic aircraft in Section 7.3 is at Mach 1.6 with eight passengers and at a range of $7408km$. Comparing the value with the result at $Ma = 1.6$ with ten passengers and a range of $7408km$ one observes that the increase in payload results in a decrease in DOC by 25.18%. The results are shown in Table B.19.

[10] Assume that the number of passengers is not a big game changer, since the aircraft is small and its size is defined.

[11] The range could nevertheless have an impact here, depending on how the utilization $u_F$ is calculated. Assume here that the utilization is a given value.

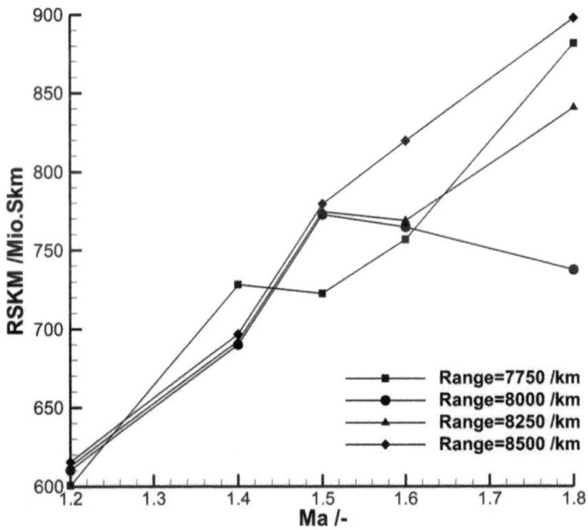


**Figure 7.12:** Parameter Variation: mach number dependence of revenue seat kilometres, $RSKM$

Take a closer look at the, in terms of DOC, most favourable combinations of Mach number and range for the given aircraft geometry. They are depicted in Table 7.7.

**Table 7.7:** Parameter variation: DOC of favourable combinations of Mach number and range for 10 passengers

| Option | $Ma/-$ | $R_F/km$ | $DOC/\frac{EURO}{SKM}$ |
|---|---|---|---|
| 1 | 1.4 | 7500 | 0.33107 |
| 2 | | 7750 | 0.33086 |
| 3 | 1.5 | 7250 | 0.33158 |
| 4 | | 7500 | 0.33162 |
| 5 | | 8000 | 0.33129 |
| 6 | | 8250 | 0.33277 |
| 7 | | 8500 | 0.33301 |
| 8 | | 8750 | 0.33390 |
| 9 | 1.6 | 8500 | 0.33439 |

As outlined in Section 7.2.3 ranges of $6640km$, $7900km$ and $9300km$ have been identified as to having promising market penetration potential. The results show that a range of $9000km$ is barely reachable with the given aircraft configuration; more precisely it is only achievable at a speed of Mach 1.5 since otherwise the maximum loadable fuel mass ($36153.kg$) would be exceeded (cf. Table B.18(h)). The combination of a range of $7750km$ and $Ma = 1.4$ (Option 2) shows the lowest DOC. Under the rough assumption that the market penetration potential and the range show a linear relationship, hence linear interpolation is applicable, the given aircraft would cover slightly below 60% of the estimated market. In order to reach a range of $7900km$ and thus according to Liebhardt et al. (2013) cover 60% of the market, a payload reduction of $279kg$ or three passengers with luggage would be necessary. This is shown in the mass-range chart given in Figure 7.13(a). However, a higher Mach number is desirable since the major argument for a SSBJ is speed, and consequently options 1 and 2 are abandoned. The most promising combination of Mach number and range is option 5. This configuration allows for a range of $8000km$, hence would cover 60% of the estimated market at a speed of Mach 1.5. Compared to option 2 no payload reduction has to be considered to reach this range. The change in DOC of 0.12% when compared to option 2 is negligible since the uncertainties within the frame of preliminary aircraft design are considered to be higher. Options 3, 4, 6, 7 and 8 do not seem reasonable when examining the range, since the next jump in marked coverage is, according to Liebhardt et al., expected to be at $9380km$ and a range of $7250km$ is too short. The increase in range and speed of option 9 would not generate a benefit in terms of market coverage, even though the increase in speed is attractive. Based on the analysed data option 5, seems to be the most suitable one for the given aircraft configuration. The mass-range chart for option 5 is given in Figure 7.13(b) without further discussion.

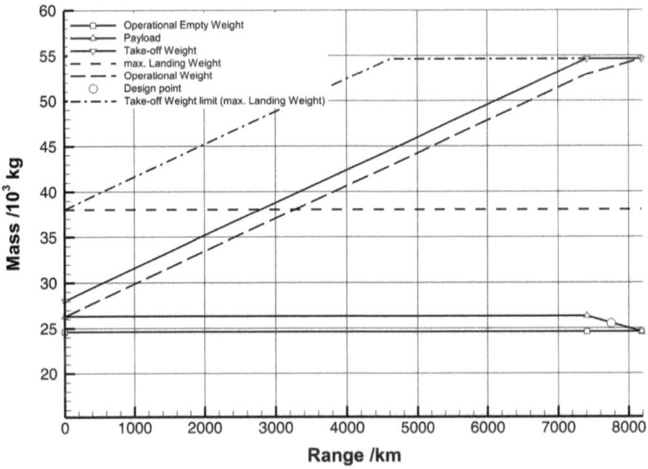


(a) $Ma = 1.4$ and $7750km$

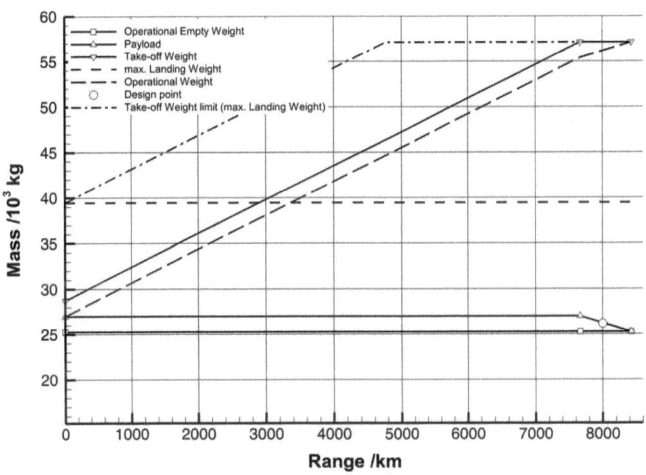


(b) $Ma = 1.5$ and $8000km$

**Figure 7.13:** Parameter variation: mass-range chart

### 7.4.2 Performance Studies for Subsonic and Supersonic Cruise Conditions

It is unlikely that the cruise phase of a supersonic business jet is solely supersonic. Overland segments will be required for commercial success and since the overland flight ban is in force, subsonic flight is necessary. Please note that the term *subsonic* in this part of the text relates to all Mach numbers below the speed of sound. The question that will be answered briefly here is, "What range can a given aircraft fly when used on missions composed from subsonic and supersonic cruise segments, hence missions differing from its design mission?" Based on the parameter variations in Section 7.4.1, the aircraft option 5 ($R = 8000km$ and $Ma = 1.5$) is selected to examine the performance on subsonic/supersonic mission profiles. The number of possible combinations of subsonic and supersonic segments is unlimited and therefore reasonable scenarios have to be selected. Liebhardt et al. (2011) examined promising routes for supersonic business jets and conclude that *three quarters of them were found to possibly have supersonic cruise percentages of 80% and more.* Based on this statement, eight mission scenarios have been developed and are given in Table 7.8. In this table, the parameters *coding* and *fraction* have to be explained. *Coding* indicates the order of segments, while *sub* is a subsonic segment and *super* a supersonic segment. For example, scenario four consists of a subsonic segment followed by a supersonic segment and again a subsonic segment. The parameter *fraction* gives the cruise percentages of each segment. Before running the analysis, some more boundary conditions have to be mentioned. The Take-Off Mass ($m_{TOM}$) of the aircraft shall correspond to its Maximum Take-Off Mass, $m_{TOM} = m_{MTOM} = 57109.kg$. Payload and mission fuel mass are $910.kg$ and $30937.2kg$, respectively. Note that the fuel mass in Table B.18(d) is the maximum fuel mass, whereas the mass used here is the fuel mass for the design mission of the aircraft, cf. Figure 7.13(b). The payload is that of the design mission. For supersonic segments, the design Mach number $Ma = 1.5$ is used, whereas for subsonic flight a value of $Ma = 0.95$ is chosen. Remember that the advantage of a SSBJ is its speed which makes it reasonable to fly at high Mach numbers when flying below $Ma = 1.0$.

Before discussing the results, remember what has been discussed in Section 7.1: Direct Operating Costs are used in this thesis as a way to quantify sensitivities. In the end, *soft factors* might be more relevant than cost!

The resultant range ($R_7$) and the Direct Operation Costs ($DOC_7$) are given in Table 7.9 and depicted in Figure 7.14. The scenarios 1 and 10 represent 100% and 0% supersonic cruise. The rectangles in the figure group the results of scenarios with a supersonic fraction of 80%, 50% and 20% in cruise.

The scenarios where the supersonic segment is flown prior to the subsonic segment (scenarios 2, 5 and 8) show better results in terms of range and DOC compared to the scenarios flown in inverse order (scenarios 3, 6 and 9). Those scenarios in which cruise flight is divided into three segments (scenarios 4 and 7) show the worst results when compared within the respective reference group. The reason for this behaviour is straightforward: the aircraft is designed to fly at supersonic speeds. The unfavourable subsonic performance is being slightly compensated by a lighter aircraft for the scenarios 2, 3 and 8, since a major part of the fuel is burned in the first segment which is the supersonic one.

**Table 7.8:** Performance studies: scenarios for subsonic and supersonic cruise conditions

| Scenario No. | Coding | Fraction /% |
|---|---|---|
| 1 | super | 100 |
| 2 | super/sub | 80/20 |
| 3 | sub/super | 20/80 |
| 4 | sub/super/sub | 10/80/10 |
| 5 | super/sub | 50/50 |
| 6 | sub/super | 50/50 |
| 7 | sub/super/sub | 25/50/25 |
| 8 | super/sub | 20/80 |
| 9 | sub/super | 80/20 |
| 10 | sub | 100 |

**Table 7.9:** Performance Studies: range and DOC for subsonic and supersonic cruise scenarios for a given aircraft (Design point $Ma = 1.5$ and $R = 8000km$)

| Scenario No. | $R_7/km$ | $DOC_7/\frac{\mathrm{EURO}}{SKM}$ |
|---|---|---|
| 1 | 8000.00 | 0.33129 |
| 2 | 7910.49 | 0.34667 |
| 3 | 7863.34 | 0.34990 |
| 4 | 7615.69 | 0.36229 |
| 5 | 7729.79 | 0.36841 |
| 6 | 7621.64 | 0.37411 |
| 7 | 7411.27 | 0.38762 |
| 8 | 7543.32 | 0.38889 |
| 9 | 7380.63 | 0.39781 |
| 10 | 7198.06 | 0.41560 |

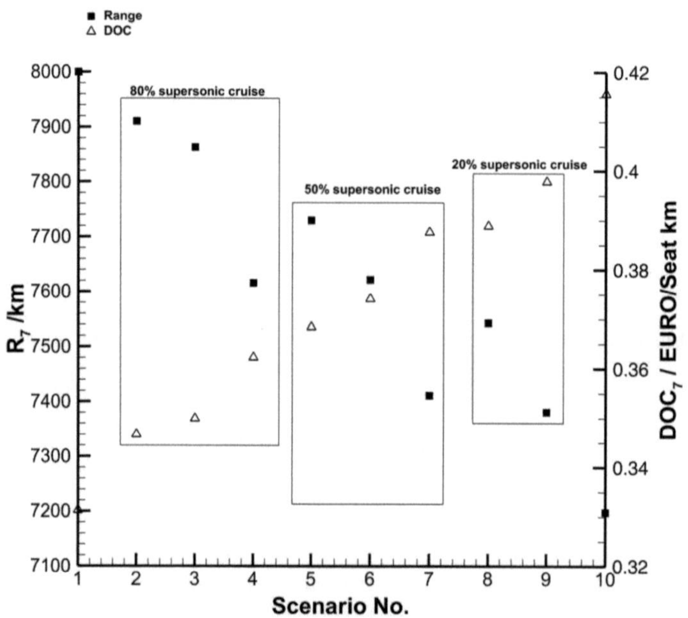


**Figure 7.14:** Performance Studies: range and DOC for subsonic and supersonic cruise scenarios for a given aircraft (Design point $Ma = 1.5$ and $R = 8000km$)

# 8 Summary and Conclusions

In this thesis, the Preliminary Aircraft Design and Optimisation tool, PrADO, is extended by methods to evaluate supersonic transport aircraft, especially supersonic business jets (SSBJ). Therefore, an overview of existing research and development projects is given before the process of conceptual and preliminary aircraft design is outlined and PrADO is introduced. The focus of this work lies in the extension of PrADO's methodology by methods for the calculation of aerodynamic coefficients, structural mass estimation and propulsion, thus providing a reliable tool to evaluate future supersonic transport aircraft in a conceptual and preliminary design context.
A chapter on fundamentals gives relevant information on theoretical aspects and references literature for further studies. Especially, the inviscid Euler equations are introduced to the preliminary design environment and information on a turbofan engine with mixed exhaust flow that is introduced into the process is given. However, the key aspect of the work presented is the development of a structural sizing algorithm to estimate structural mass based on a newly introduced six degree of freedom finite element model. Therefore, the author's ideas and suggestions are developed. How the discussed theories and ideas are put to use is explained in a chapter on modelling aspects. The geometry data for all models is directly taken from PrADO.
A challenge in preliminary aircraft design is the verification or validation of the tool chain due to the lack of data. This proves true especially for future and unconventional aircraft like a SSBJ. That said, the developed methods are being verified in a two-step approach. Firstly, the aerodynamic method as well as the structural sizing process are being verified in a stand-alone manner. The verification of the aerodynamic data is solely based on its expected qualitative distribution, since data from higher order methods was not available for the time being. Nonetheless, the results compare well with the expected behaviour. The structural sizing process is verified based on an example from literature. The results of the developed sizing algorithm show extraordinary agreement with the reference data. The second step of the verification process applies the complete aircraft design process on an example from literature. More precisely, an aircraft from the European research project HISAC is taken as a reference. The results of PrADO compare well with the available global aircraft data.
An excursion into the field of temperature effects in supersonic flight is provided after the verification example. This seemed necessary, since no relevant literature is found on the topic with regard to conceptual and preliminary aircraft design. Therefore, an analysis model is set up and used to examine the general temperature distribution on a supersonic aircraft. The results are translated into helpful information on the material selection process in the stage of aircraft pre-design.

Finally, the tool chain is applied to analyse a supersonic business jet and the results are presented. Based on these results, a parameter study is conducted. For this study, the aerodynamic shape of the aircraft is kept fix to save computational time. Instead, the operational parameters, more precisely the design point as a combination of cruise Mach number and design range is varied. Global design parameters, i.e. masses and static thrust, show sensitivity to such variations. The operating empty mass and static thrust are more sensitive to Mach number variations than to a change in range. The Maximum take-off mass shows more sensitivity to changes of the design range than to a change of the Mach number. Even though standard Direct Operating Cost (DOC) models should be used with caution when it comes to business travel, one of them is used as an evaluation criterion. The sensitivity of DOC values to changes in Mach number and range is shown and follows the general expectations.
It is normal that the aircraft will not solely be operated on its design mission but on different missions. For a supersonic aircraft, this mission can consist of segments below the speed of sound. Therefore, a mission analysis is performed to show the aircraft's behaviour on missions differing from the design mission. Resulting from the parameter variations, the aircraft used for this analysis does have a design Mach number of $Ma = 1.5$ and a design range of $8000 km$. As expected, the aircraft shows poorer performance when operated away from its design point. It is shown that a higher fraction of supersonic cruise results in a better performance than a low portion and that the order of subsonic and supersonic segments influences the performance significantly.
The work presented in this thesis motivates further research on supersonic business jets and the extension of PrADO. The results suggest, that research on Direct Operating Cost models for supersonic travel and business travel should be conducted and implemented into the software. Regarding the optimization of a given aircraft, the decrease of computational time of the aerodynamic analysis process would provide a major enhancement.

# References


Adelman, H. M., Haftka, R. T., & Tsach, U. (1980). *Application of Fully Stressed Design Procedures to Redundant and Non-Isotropic Structures.* Technical Report NASA-TM-81842, NASA, Langley Research Center.

Aerion (2013). *http://aerioncorp.com.* Accessed on: December 4th, 2013.

Anderson, J. D. (1995). *Computational Fluid Dynamics: the Basics with Applications.* McGraw-Hill, New York.

Anemaat, W., Kaushik, B., Hale, R., & Ramabadran, N. (2007). *AAARaven: Knowledge-Based Aircraft Conceptual and Preliminary Design.* In *48th AIAA/ASME/ASCE/AHS/ASC Structures, Structural Dynamics, and Materials Conference, Honolulu, HI*, number AIAA 2007-2291. AIAA.

Anemaat, W. A. & Kaushik, B. (2011). *Geometry Design Assistant for Airplane Preliminary Design.* In *49th AIAA Aerospace Sciences Meeting including the New Horizons Forum and Aerospace Exposition*, number AIAA 2011-162.

Anon (2006). *Procedures for Air Navigation Services: Aircraft Operations - Flight Procedures* (5 ed.)., volume 1. International Civil Aviation Organization (ICAO).

Anon (2010). *Jet A-1 Kerosine Safty Data Sheet.* Technical report, Air BP, BP Middle East.

Anon (2011). *MSC NASTRAN 2012 Linear Static Analysis User's Guide.* MacNeal-Schwendler Corporation.

Anon (2014a). *http://www.merriam-webster.com/.* Accessed on: April 16th, 2014.

Anon (2014b). *TAU-Code User Guide.*

Antoine, N., Kroo, I., Willcox, K., & Barter, G. (2005). *Framework for Aircraft Conceptual Design and Environmental Performance Studies. AIAA Journal, 43*(10), 2100.

Aronstein, D. & Schueler, K. (2004). *Conceptual Design of a Sonic Boom Constrained Supersonic Business Aircraft.* In *42nd AIAA Aerospace Sciences Meeting and Exhibition, Reno*, (pp. 2004–0697).

Bathe, K.-J. (2002). *Finite-Elemente-Methoden* (2 ed.). Springer Berlin.

Bekemeyer, P. (2014). *Voruntersuchungen zum Einfluss der Machzahl auf die Strukturanalyse von Überschallflugzeugen im Rahmen des Gesamtentwurfs (Preliminary Assessment of the Influence of the Mach Number on Structural Analysis of Supersonic Aircraft in the Framework of Aircraft Pre-Design).* Master's thesis, TU-Braunschweig.

Boppe, C. W. (1992). *Aircraft Drag Analysis Methods.* In P. Sacher (Ed.), *Special Course on Engineering Methods in Aerodynamic Analysis and Design of Aircraft*, number ISBN-92-835-0652-9, AGARD-R-783 chapter 7. Advisory Group for Aerospace Research and Development of the North Atlantic Treaty Organisation (NATO).

Brezillon, J., Carrier, G., & Laban, M. (2011). *Multidisciplinary Optimization of Supersonic Aircraft Including Low-Boom Considerations. Journal of Mechanical Design, 133*(10), 105001–105009.

Bräunling, W. J. G. (2004). *Flugzeugtriebwerke: Grundlagen, Aero-Thermodynamik, Kreisprozesse, Thermische Turbomaschinen, Komponenten-und Emissionen.* Springer.

Christensen, R. M. (2013). *The World Wide Failure Exercise II Examination of Results. Journal of Reinforced Plastics and Composites, 32*(21), 1668–1672.

Chudoba, B., Coleman, G., Czysz, P., & Butler, C. (2009). *Feasibility Study of a Supersonic Business Jet based on the Learjet Airframe. Aeronautical Journal, 113*(1146), 517.

Chudoba, B., Coleman, G., Oza, A., & Czysz, P. (2008). *What Price Supersonic Speed? A Design Anatomy of Supersonic Transportation Part 1. Aeronautical Journal, 112*(1129), 141–151.

Davies, R. (1998). *Supersonic (Airliner) Non-Sense - A Case Study in Applied Market Research.* Paladwr Press.

Deremaux, Y. (2009). *Executive Public Summary: Final Aircraft Configuration Families.* Technical Report D5.23 HISAC T-5-36-1, Dassault Aviation.

Donaldson, B. K. (1993). *Analysis of Aircraft Structures: an Introduction.* Number ISBN 0-07-112591-4. McGraw-Hill.

Eisfeld, B. (2002). *Die Reynolds-Spannungsgleichungen für kompressible Strömungen - Herleitung und Zusammenhänge.* Technical Report IB 124-2002/36, Deutsches Zentrum für Luft- und Raumfahrt e.V.

Erickson, L. (1990). *Panel Methods: an Introduction.* Technical Report 2995, NASA.

Fiecke, D. (1958). *Die Bestimmung der Flugzeugpolaren für Entwurfszwecke - zweiter Teil: Anwendung.* In *Technische und volkswirtschaftliche Berichte des Wirtschafts- und Verkehrsministeriums Nordrhein-Westfalen*, number 44. Wirtschafts- und Verkehrsministerium Nordrhein-Westfalen.

Flack, R. D. (2005). *Fundamentals of Jet Propulsion with Applications.* Cambridge University Press, New York.

Fornasier, L. (1985). *Linearized Potential Flow Analysis of Complex Aircraft Configurations by HISSS, a Higher Order Panel Method.* In *23rd AIAA Aerospace Sciences Meeting and Exhibition, Reno*, volume 14. AIAA.

Grewe, V. (2007). *Climate Impact of a Potential Supersonic Fleet.* In *1st CEAS European Air and Space Conference,*, number CEAS-2007-421. CEAS. VIP.

Gudmundsson, S. (2013). *General Aviation Aircraft Design: Applied Methods and Procedures.* Number ISBN 978-0-12-397308-5. Butterworth Heinemann.

Gur, O., Mason, W. H., & Schetz, J. A. (2009). *Full Configuration Drag Estimation.* In *27th AIAA Applied Aerodynamics Conference.*

Haftka, R. T. & Gürdal, Z. (1992). *Elements of Structural Optimization* (3 ed.). Number ISBN 0-7923-1504-9. Kluwer Academic Publishers, Dordrecht, The Netherlands.

Hansen, L. U., Heinze, W., & Horst, P. (2008). *Blended Wing Body Structures in Multidisciplinary Pre-Design. Structural and Multidisciplinary Optimization, 36*(1), 93–106. IFL.

Hart-Smith, L. (1998). *Predictions of the Original and Truncated Maximum-Strain Failure Models for Certain Fibrous Composite Laminates. Composites Science and Technology, 58*(7), 1151 – 1178.

Harzheim, L. (2008). *Strukturoptimierung: Grundlagen und Anwendungen.* Harri Deutsch Verlag.

Haupt, M., Niesner, R., Unger, R., & Horst, P. (2005). *Coupling Techniques for Thermal and Mechanical Fluid-Structure-Interactions in Aeronautics. PAMM, 5*(1), 19–22.

Heinze, W. (1994). *Ein Beitrag zur quantitativen Analyse der technischen und wirtschaftlichen Auslegungsgrenzen verschiedener Flugzeugkonzepte für den Transport großer Nutzlasten.* PhD thesis, TU Braunschweig. ZLR-Forschungsbericht 94-01.

Heinze, W. (2005). *Entwerfen von Verkehrsflugzeugen II - Vorlesungsmanuskript.* Number IFL-02/05. Institut für Flugzeugbau und Leichtbau, TU-Braunschweig.

Heinze, W. (2007). *Entwerfen von Verkehrsflugzeugen I - Vorlesungsmanuskript.* Number IFL-01/07. Institut für Flugzeugbau und Leichtbau, TU-Braunschweig.

Heinze, W., Österheld, C. M., & Horst, P. (2001). *Multidisziplinäres Flugzeugentwurfsverfahren PrADO - Programmentwurf und Anwendung im Rahmen von Flugzeug-Konzeptstudien.* In *Jahrbuch der DGLR-Jahrestagung 2001 in Hamburg.* DGLR.

Henne, P. A. (2005). *Case for Small Supersonic Civil Aircraft. Journal of aircraft, 42*(3), 765–774.

Herrmann, U. & Laban, M. (2007). *Multi-Disciplinary Analysis and Optimisation Applied to Supersonic Aircraft; Part 2: Application and Results.* In *48th AIAA/ASME/ASCE/AHS/ASC Structures, Structural Dynamics, and Materials Conference, Honolulu, HI,* number 2007-1858. AIAA, AIAA.

Hoerner, S. (1965). *Fluid-Dynamic Drag: Practical Information on Aerodynamic Drag and Hydrodynamic Resistance.* Hoerner Fluid Dynamics.

Horstmann, K. (1987). *Ein Mehrfach-Traglinienverfahren und seine Verwendung für Entwurf und Nachrechnung nichtplanarer Flügelanordnungen.* PhD thesis, Deutsche Forschungs-und Versuchsanstalt für Luft-und Raumfahrt (DFVLR).

Howe, D. (2000). *Aircraft Conceptual Design Synthesis.* Professional Engineering Publishing.

Hürlimann, F., Kelm, R., Dugas, M., & Kress, G. (2012). *Investigation of Local Load Introduction Methods in Aircraft Pre-Design. Aerospace Science and Technology, 21*(1), 31 – 40.

Hürlimann, F., Kelm, R., Dugas, M., Oltmann, K., & Kress, G. (2011). *Mass Estimation of Transport Aircraft Wingbox Structures with a CAD/CAE-Based Multidisciplinary Process. Aerospace Science and Technology, 15*(4), 323 – 333.

Huda, Z. & Edi, P. (2013). *Materials selection in design of structures and engines of supersonic aircrafts: A review. Materials & Design, 46*(0), 552 – 560.

Kaddour, A. S. & Hinton, M. J. (2013). *Maturity of 3D Failure Criteria for Fibre-Reinforced Composites: Comparison Between Theories and Experiments: Part B of WWFE-II. Journal of Composite Materials, 47*(6-7), 925–966.

Kaddour, A. S., Hinton, M. J., & Soden, P. D. (2013). *The Background to the Third World-Wide Failure Exercise. Journal of Composite Materials, 47*(20-21), 2417–2426.

Kaenel, R. v., Rizzi, A., Oppelstrup, J., Goetzendorf-Grabowski, T., Ghoreyshi, M., Cavagna, L., & Bérard, A. (2008). *Ceasiom: Simulating stability and control with cfd/csm in aircraft conceptual design.* In *26th International Congress of the Aeronautical Sciences, ICAS.*

Kopenhagen, W. (1996). *Das große Flugzeugtypenbuch* (7 ed.). Motorbuch Verlag.

Kreuzer, H. (2002). *Moderne Verkehrsflugzeuge und Business Jets.* Number ISBN 3-9805934-0-1. Am Start.

La Rocca, G. & Tooren, M. J. L. v. (2007). *A Knowledge Based Engineering Approach to Support Automatic Generation of FE Models in Aircraft Design.* In *45th AIAA Aerospace Sciences Meeting and Exhibit*, (pp. 2007–0967). TU - Delft.

Laban, M. & Herrmann, U. (2007). *Multi-Disciplinary Analysis and Optimisation Applied to Supersonic Aircraft; Part 1: Analysis Tools.* In *48th AIAA/ASME/ASCE/AHS/ASC Structures, Structural Dynamics, and Materials Conference, Honolulu, HI*, number 2007-1857. AIAA, AIAA.

Lecheler, S. (2009). *Numerische Strömungsberechnung: Schneller Einstieg durch ausführliche praxisrelevante Beispiele.* Vieweg+ Teubner Verlag.

Li, Q., Steven, G., & Xie, Y. (2001). *A Simple Checkerboard Suppression Algorithm for Evolutionary Structural Optimization. Structural and Multidisciplinary Optimization, 22*(3), 230–239.

Liebhardt, B., F., L., & K., D. (2013). *Supersonic Diversion - Assessment of Great-Circle versus Sonic Boom-Restricted Flight Routing.* In *AIAA Aviation Technology, Integration, and Operations (ATIO).* AIAA.

Liebhardt, B. & Lütjens, K. (2011). *An Analyses of the Market Environment for Supersonic Business Jets.* In *Deutscher Luft- und Raumfahrtkongress 2011 - Manuskripte*, number 241457, (pp. 617 – 627). DGLR e.V., DGLR e.V.

Liebhardt, B., Lütjens, K., & Gollnick, V. (2011). *Estimation of the Market Potential for Supersonic Airliners via Analysis of the Global Premium Ticket Market.* In *11th AIAA Aviation Technology, Integration, and Operations (ATIO) Conference, Virginia Beach, VA, USA*, number AIAA 2011-6806. AIAA, AIAA.

Liu, K.-S. & Tsai, S. W. (1998). *A Progressive Quadratic Failure Criterion for a Laminate. Composites Science and Technology, 58*(7), 1023 – 1032.

LTH (2006). *Luftfahrttechnisches Handbuch.* IABG mbH, LTH-Koordinierungsstelle.

Mack, R. (2003). *A Supersonic Business-Jet Concept Designed for Low Sonic Boom.* Technical report, NASA Langley Research Center, TM-2003-212435, Hampton, VA.

Malone, B. & Myklebust, A. (1996). *ACSYNT: Commercialization Success.* Technical Report 965568, SAE International.

Mason, W., Knill, D., Giunta, A., Grossman, B., Watson, L., & Haftka, R. (1998). *Getting the full benefits of CFD in conceptual design.* In *16th AIAA Applied Aerodynamics Conference, Albuquerque*, number AIAA 98-2513. AIAA.

Mattingly, J., Heiser, W., & Daley, D. (1987). *Aircraft Engine Design.* American Institute of Aeronautics and Astronautics, Inc.

Mavris, D. N. & Hayden, W. T. (1996). *Formulation of an IPPD Methodology for the Design of a Supersonic Business Jet.* Technical report, Georgia Institute of Technology.

Menter, F. R. (1994). *Two-Equation Eddy-Viscosity Turbulence Models for Engineering Applications. AIAA journal, 32*(8), 1598–1605.

Michell, A. G. M. (1904). *The Limits of Economy of Material in Frame-Structures. Philosophical Magazine, 8(47)*, 589–597.

MIL-HDBK-5J (2003). *Metallic Materials and Elements for Aerospace Vehicle Structures, MIL-HDBK-5J.* U.S. Department of Defense , Washington, D.C.

Mukhopadhyay, V. (2007). *A Conceptual Aerospace Vehicle Structural System Modeling, Analysis and Design Process.* In *48th AIAA/ASME/ASCE/AHS/ASC Structures, Structural Dynamics, and Materials Conference, Honolulu, HI*, number AIAA 2007-2372. AIAA.

Nagel, B., Böhnke, D., Gollnick, V., Schmollgruber, P., Rizzi, A., La Rocca, G., & Alonso, J. J. (2012). *Communication in Aircraft Design: Can we Establish a Common Language?* In *28th International Congress of the Aeronautical Sciences ICAS, Brisbane, Australia.*

Nickol, C. L. & A., M. L. (2009). *Hybrid Wing Body Configuration System Studies.* In *47th AIAA Aerospace Sciences Meeting including The New Horizons Forum and Aerospace Exposition*, number AIAA 2009-931. AIAAA.

Niu, M. C. Y. (1999). *Airframe Structural Design: Practical Design Information and Data on Aircraft Structures.* Conmilit Press Limited.

Österheld, C. M. (2003). *Physikalisch begründete Analyseverfahren im integrierten multidisziplinären Flugzeugvorentwurf.* PhD thesis, TU-Braunschweig, Zentrum für Luft- und Raumfahrt.

Österheld, C. M., Heinze, W., & Horst, P. (2001). *Preliminary Design of a Blended Wing Body Configuration using the Design Tool PrADO.* In *Proceedings der CEAS Conference on Multidisciplinary Aircraft Design and Optimisation.*

Paulson, M. (2007). *Putting a Q into Supersonic Flight. Professional Pilot Magazine, 41*(3), 58–62.

Paulson, M. (2013). *Time is Right to Go Forward with Supersonic Business Jets. Professional Pilot Magazine*, *47*(3), 52–54.

Puck, A. (1996). *Festigkeitsanalyse von Faser-Matrix-Laminaten: Modelle für die Praxis.* Carl Hanser Verlag, München, Wien.

Raymer, D. P. (2006). *Aircraft design: A Conceptual Approach* (4 ed.). American Institute of Aeronautics and Astronautics.

Raymer, D. P. (2011). *Conceptual Design Modeling in the RDS-Professional Aircraft Design Software.* In *49th AIAA Aerospace Sciences Meeting including the New Horizons Forum and Aerospace Exposition*, number AIAA 2011-161. AIAA.

Rentema, D. (2004). *AIDA. Artificial Intelligence Supported Conceptual Design of Aircraft.* PhD thesis, TU Delft.

Rieke, J. (2013). *Bewertung von CFK-Strukturen in einem multidisziplinären Entwurfsansatz für Verkehrsflugzeuge.* Number ISBN 978-3-95404-368-2. Culiver Verlag Göttingen.

Risse, K., Anton, E., Lammering, T., Franz, K., & Hoernschemeyer, R. (2012). *An Integrated Environment for Preliminary Aircraft Design and Optimization.* In *53rd AIAA Structures, Structural Dynamics and Materials Conference, Honolulu, HI*, number AIAA 2012-1675.

Rizzi, A., Eliasson, P., Goetzendorf-Grabowski, T., Vos, J., Zhang, M., & Richardson, T. (2011). *Design of a canard configured TransCruiser using CEASIOM. Progress in Aerospace Sciences*, *47*(8), 695–705.

Roskam, J. (1989a). *Airplane Design: Component Weight Estimation*, volume 5. DARcorporation.

Roskam, J. (1989b). *Airplane Design: Layout Design of Cockpit, Fuselage, Wing and Empennage: Cutaways and Inboard Profiles*, volume 3. DARcorporation.

Roskam, J. (1990a). *Airplane Design: Airplane Cost Estimation: Design, Development, Manufacturing and Operating*, volume 8. DARcorporation.

Roskam, J. (1990b). *Airplane Design: Preliminary Calculation of Aerodynamic, Thrust and Power Characteristics*, volume 6. DARcorporation.

Roskam, J. (1991). *Airplane Design: Determination of Stability, Control and Performance Characteristics*, volume 7. DARcorporation.

Roskam, J. (1997a). *Airplane Design: Layout of Landing Gear and Systems*, volume 4. DARcorporation.

Roskam, J. (1997b). *Airplane Design: Preliminary Configuration Design and Integration of the Propulsion System*, volume 2. DARcorporation.

Roskam, J. (1997c). *Airplane Design: Preliminary Sizing of Airplanes*, volume 1. DARcorporation.

SAI (2013). *http://www.sai-qsstx.com.* Accessed on: December 4th, 2013.

Schlichting, H. (1982). *Grenzschicht-Theorie* (8th ed.). Wissenschaftliche Bücherei. Braun, Karlsruhe.

Schlichting, H. & Truckenbrodt, E. (2001). *Aerodynamik des Flugzeuges - Band 1* (3 ed.). Springer Berlin.

Schürmann, H. (2008). *Konstruieren mit Faser-Kunststoff-Verbunden* (2. bearbeitete und erweiterte Auflage ed.). Springer.

Schuermann, M., Gaffuri, M., & Horst, P. (2013). *Multidisciplinary Pre-Design of Supersonic Aircraft.* In *Deutscher Luft- und Raumfahrtkongress 2013 - Manuskripte*, number 301163. DGLR, (Section 1, 2, 4-7 by Schuermann).

Schuermann, M., Gaffuri, M., & Horst, P. (2014). *Extensions to Aircraft Pre-Design for Supersonic Aircraft.* In *52nd Aerospace Sciences Meeting and Exhibition, National Harbor, MD*, number AIAA 2014-0184. AIAA, (Section I-III, V-VIII by Schuermann).

Schuermann, M., Gaffuri, M., & Horst, P. (2015). *Multidisciplinary Pre-Design of Supersonic Aircraft. CEAS Aeronautical Journal*, *6*(2), 207–216, (Section 1–3, 5–8 by Schuermann).

Schumacher, A. (2005). *Optimierung mechanischer Strukturen: Grundlagen und industrielle Anwendungen.* Number ISBN 3-540-21887-4. Springer.

Schwamborn, D., Gerhold, T., & Heinrich, R. (2006). *The DLR TAU-Code: Recent Applications in Research and Industry.* In *European conference on computational fluid dynamics, ECCOMAS CFD.*

Seeckt, K. (2011). *Application of PreSTo: Aircraft Preliminary Sizing and Data Export to CEASIOM.* Technical report, Department of Aeronautics, KTH, Sweden.

Simmons, F. & Freund, D. (2005). *Morphing Concept for Quiet Supersonic Jet Boom Mitigation.* In *Proceedings of the 43rd AIAA Aerospace Sciences Meeting and Exhibit, American Institute for Aeronautics and Astronautics.* Gulfstream Aerospace Corporation, Savannah, GA 31402-2206.

Simos, D. (2006). *Piano Presentation at CAEP.* In *ICAO CAEP WG2 TG2 Rome.*

Soden, P. D., Kaddour, A. S., & Hinton, M. J. (2004). *Recommendations for Designers and Researchers Resulting from the World-Wide Failure Exercise. Composites Science and Technology*, *64*(3-4), 589 – 604.

Stephenson, W., Veley, D. E., & Hill, S. (2007). *Composite Vehicle Design Environment.* In *48th AIAA/ASME/ASCE/AHS/ASC Structures, Structural Dynamics, and Materials Conference, Honolulu, HI*, number AIAA 2007-2371. AIAA.

Stoufflet, B., de Saint Martin, P., Deremaux, Y., & Négrier, J. (2008). *Design of a Small Supersonic Transport Aircraft with High Environmental Constraints.* In *ICAS Proceedings*, number ISBN 0-9533991-9-2. ICAS.

Tidd, D., Strash, D., Epstein, B., Luntz, A., Nachshon, A., & Rubin, T. (1992). *Multigrid Euler Calculations Over Complete Aircraft. Journal of Aircraft, 29*(6), 1080–1085.

Torenbeek, E. (1982). *Synthesis of Subsonic Airplane Design.* Kluwer Academic Publisher.

Torenbeek, E. (2013). *Advanced Aircraft Design: Conceptual Design, Analysis and Optimization of Subsonic Civil Airplanes.* Number ISBN 9781119969303. John Wiley & Sons, Inc. New York, NY, USA.

Torenbeek, E., E., J., & Laban, M. (2004). *Conceptual Design and Analysis of a Mach 1.6 Airliner.* In *10th AIAA/ISSMO Multidisciplinary Analysis and Optimization Conference, Albany, NY, USA*, number AIAA 2004-4541. AIAA.

Tétrault, P.-A. (2000). *Numerical Prediction of the Interference Drag of a Streamlined Strut Intersecting a Surface in Transonic Flow.* PhD thesis, Virginia Polytechnic Institute and State University.

Weiss, T. & Heinze, W. (2013). *Multidisciplinary Design of CESTOL Aircraft with Powered Lift System.* In *SFB 880 - Fundamentals of High-Lift for Future Commercial Aircraft*, number 2013-03, (pp. 217–230).

Werner-Westphal, C., Heinze, W., & Horst, P. (2008a). *Multidisciplinary Integrated Preliminary Design Applied to Unconventional Aircraft Configurations. Journal of Aircraft, 45*, 581–590.

Werner-Westphal, C., Heinze, W., & Horst, P. (2008b). *Structural Sizing for an Unconventional, Environment-Friendly Aircraft Configuration within Integrated Conceptual Design. Aerospace Science and Technology, 12*, 184–194.

Wesseling, P. (2005). *Principles of Computational Fluid Dynamics*, volume 29. Springer.

White, F. M. (2005). *Viscous Fluid Flow* (3 ed.). Number ISBN 9780071244930. McGraw-Hill.

Wilcox, D. C. (1998). *Turbulence Modeling for CFD* (2 ed.). La Canada, CA: DCW industries.

Wiley, J. (2007). *The Super-Slow Emergence of Supersonic. Business and Commercial Aviation, 101*(3), 48.

Xie, X. & C., H. (1999). *A New Numerical Design Tool for Concept Evaluation of Propeller Aircraft. Aircraft Design, 2*(3), 147 – 165.

Zienkiewicz, O. C. & Taylor, R. L. (1989). *The Finite Element Method*, volume 1. McGraw-Hill.

# A Figures

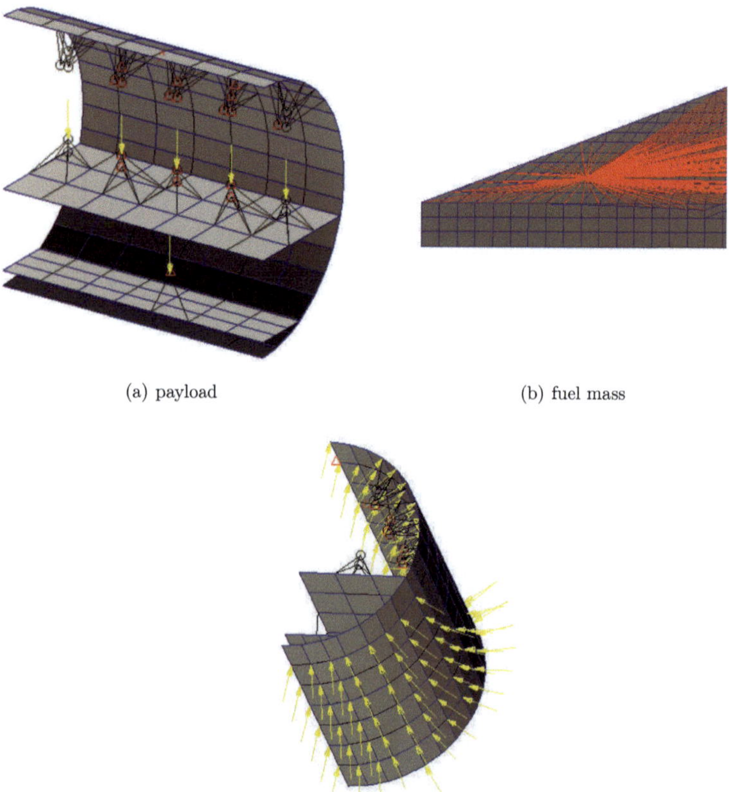

(a) payload

(b) fuel mass

(c) aerodynamic forces

**Figure A.1:** Introduction of payload, fuel masses and aerodynamic forces into the finite element model

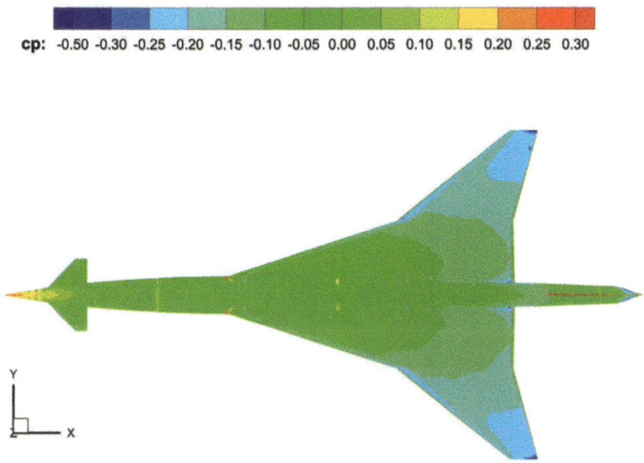


(a) Top view

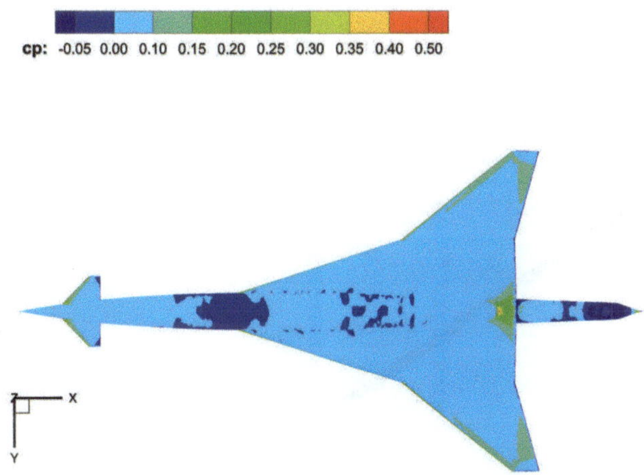


(b) Bottom view

**Figure A.2:** Verification of aerodynamics: distribution of pressure coefficient for $Ma = 1.6$ and $\alpha = 2.5deg$

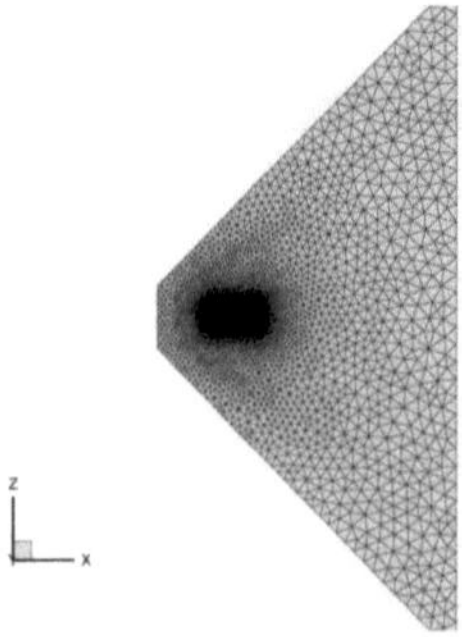

(a) Mesh at symmetry plane and farfield

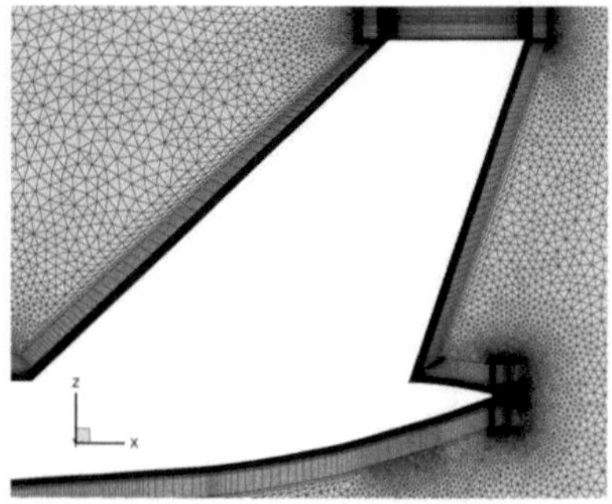

(b) Mesh at the empennage at symmetry plane

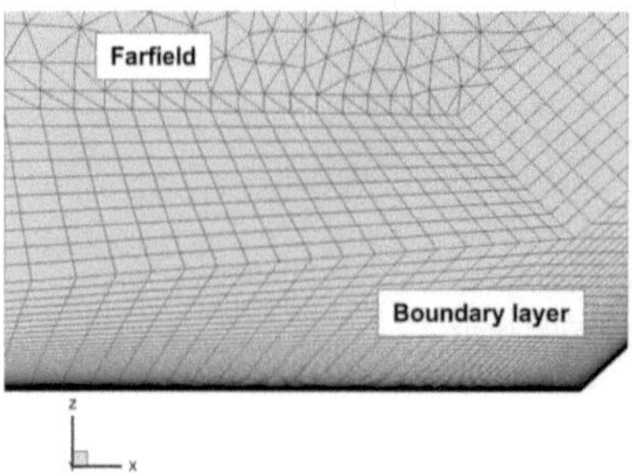


(c) Boundary layer

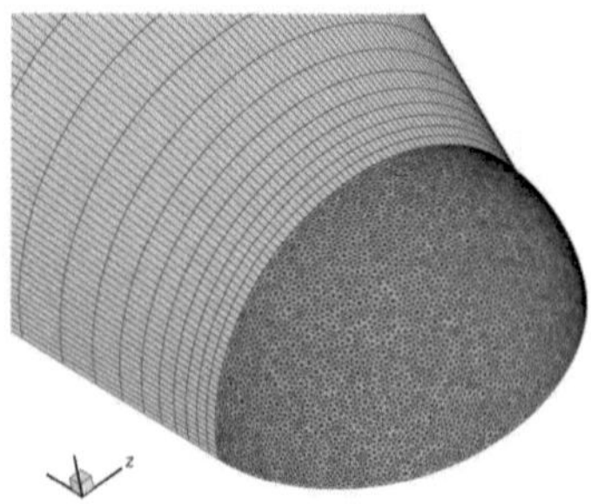

(d) Unstructured mesh at the fuselage's nose

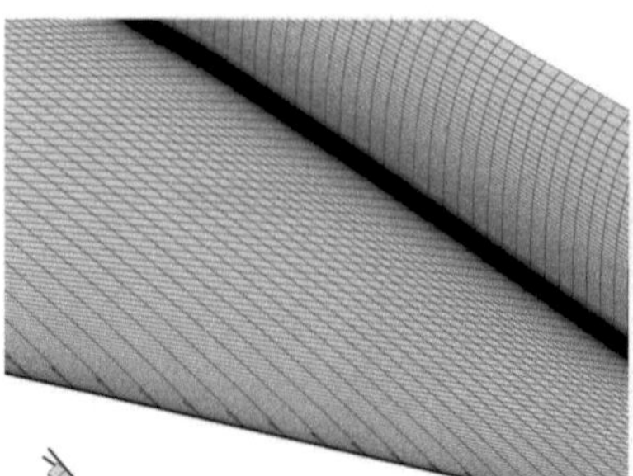

(e) Structured mesh of the wing

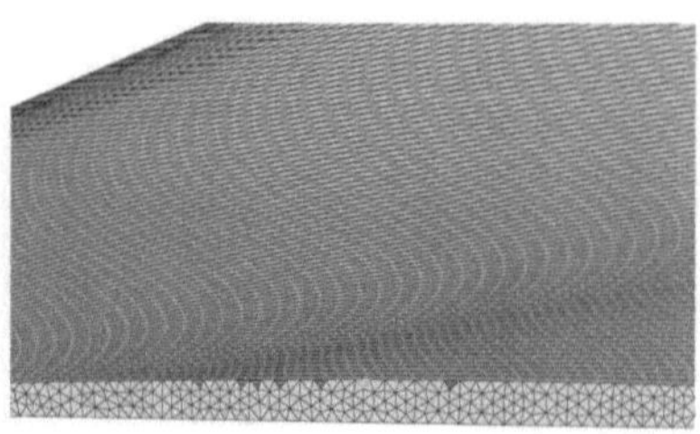

(f) Unstructured mesh at the wing tip

**Figure A.3:** Temperature analysis: CFD grid at various positions

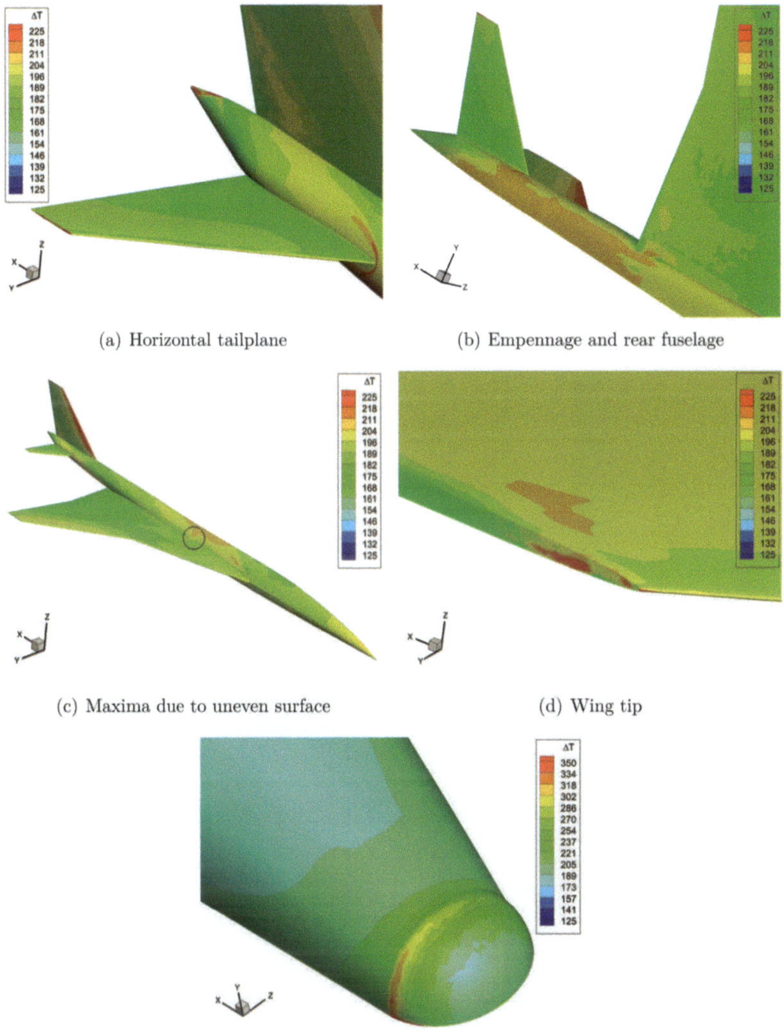


(a) Horizontal tailplane

(b) Empennage and rear fuselage

(c) Maxima due to uneven surface

(d) Wing tip

(e) Fuselage nose

**Figure A.4:** Temperature analysis: local maxima

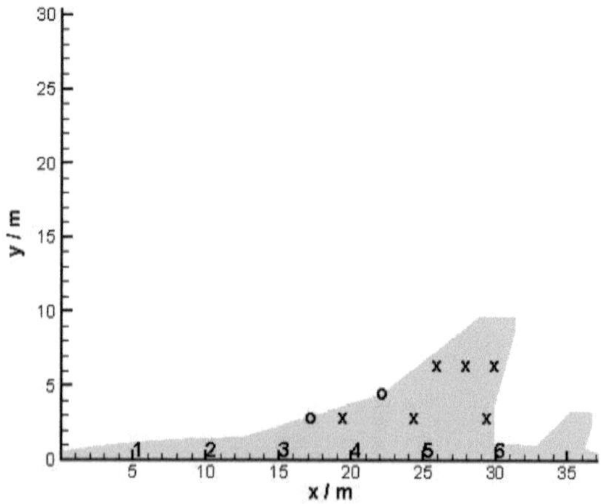


**Figure A.5:** Temperature analysis: control points
"‘x"’=leading edge, "‘o"’=wing surface, "‘1-6"’=fuselage

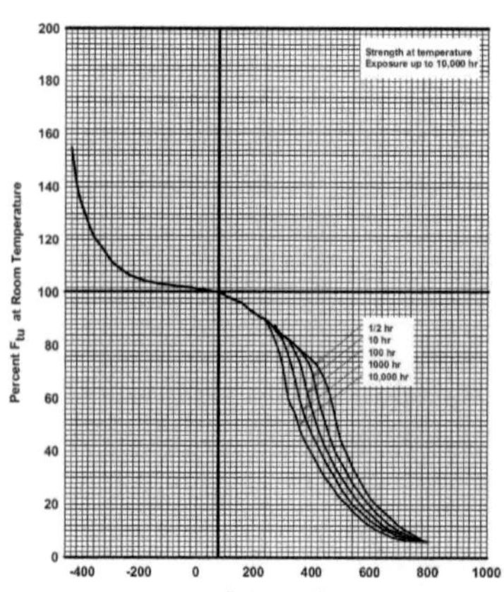


(a) Fracture toughness

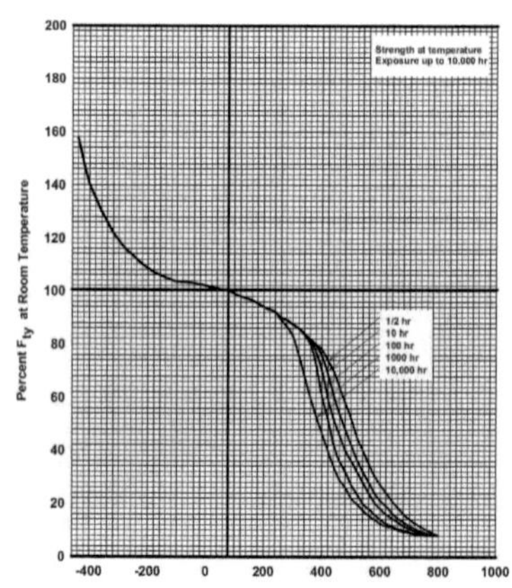


(b) Yield strength

**Figure A.6:** Temperature analysis: dependency of fracture toughness and yield strength for Al2024-T3, T351, T4 (MIL-HDBK-5J (2003))

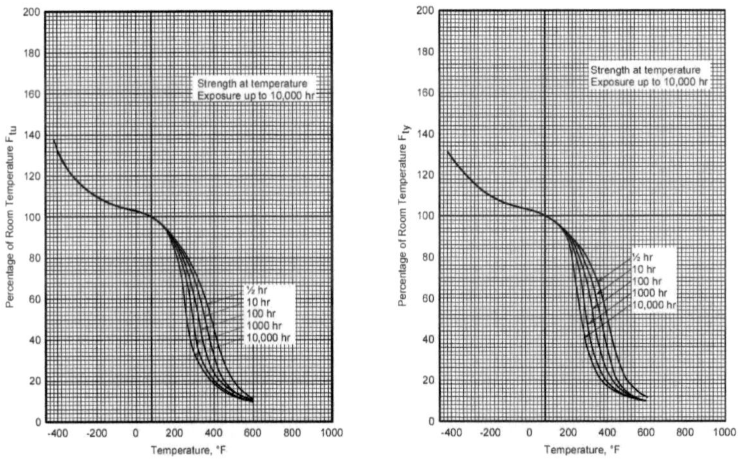


(a) Fracture toughness

(b) Yield strength

**Figure A.7:** Temperature analysis: dependency of fracture toughness and yield strength for Al7075 (MIL-HDBK-5J (2003))

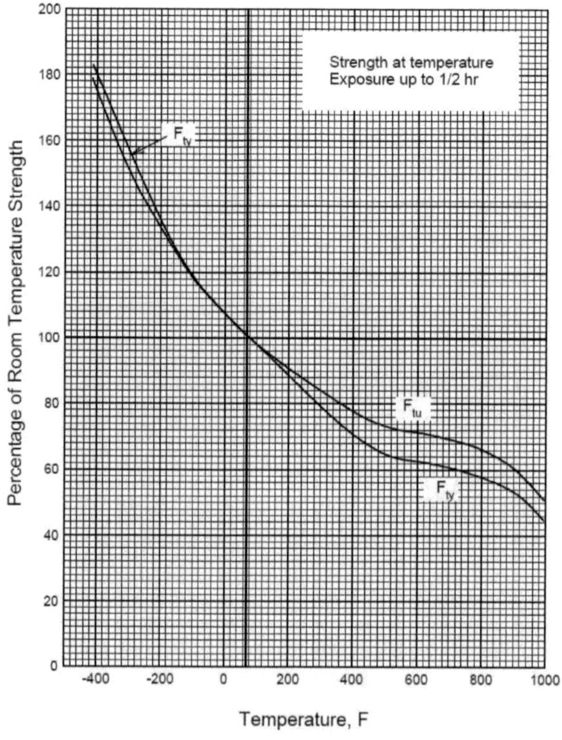


**Figure A.8:** Temperature analysis: dependency of fracture toughness and yield strength for Ti6AL-4V (MIL-HDBK-5J (2003))

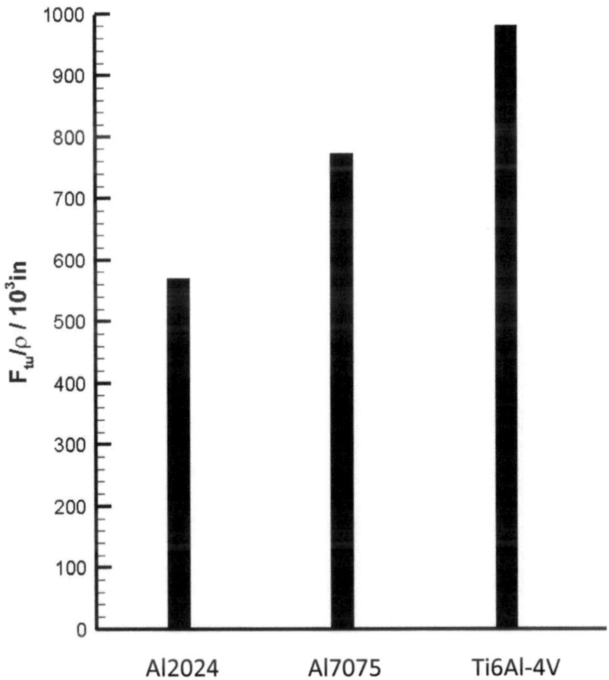


**Figure A.9:** Ultimate strength to density ratio for various materials

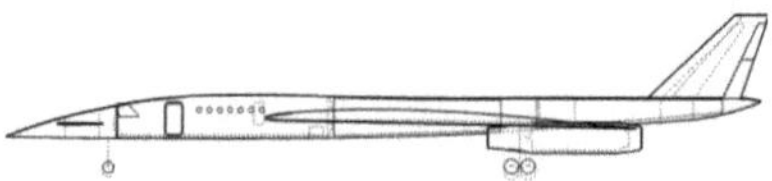

(a) Portside view

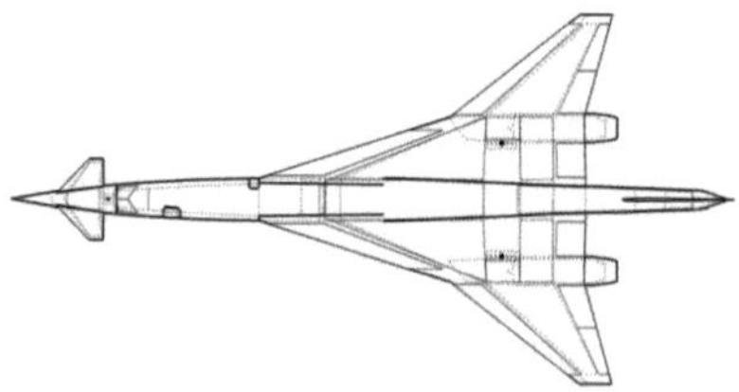

(b) Top view

**Figure A.10:** Basic aircraft: 2D view

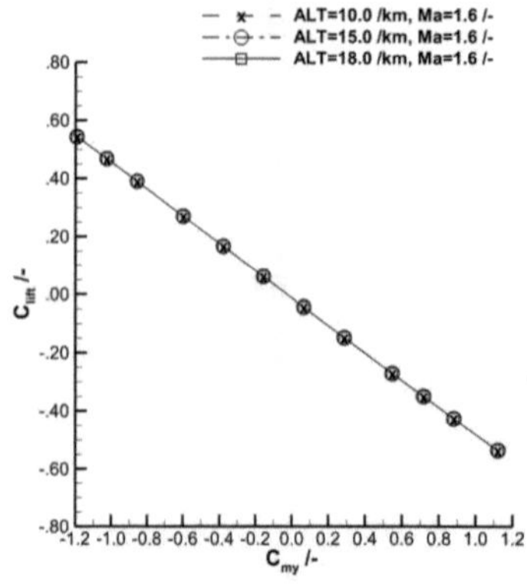


(a) $C_{lift}$ over $C_{my}$

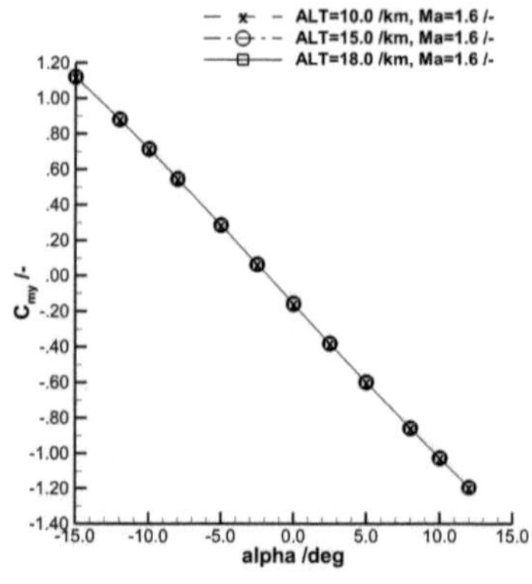


(b) $C_{my}$ over angle of attack, *alpha*

**Figure A.11:** Basic aircraft: aerodynamic pitching moment coefficient

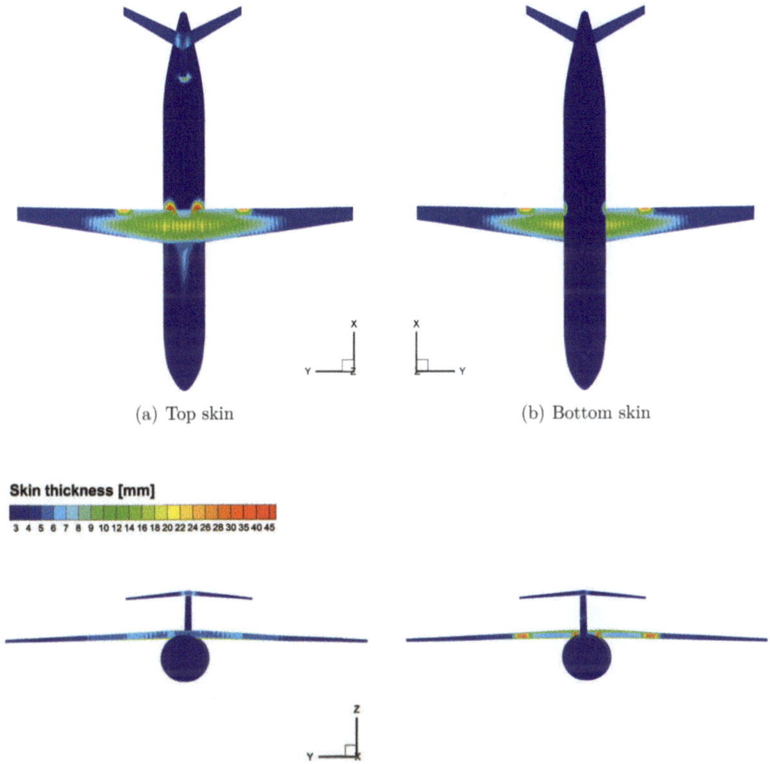


(a) Top skin

(b) Bottom skin

(c) Front view

(d) Rear view

**Figure A.12:** SFB880 aircraft: skin thickness obtained with SSM

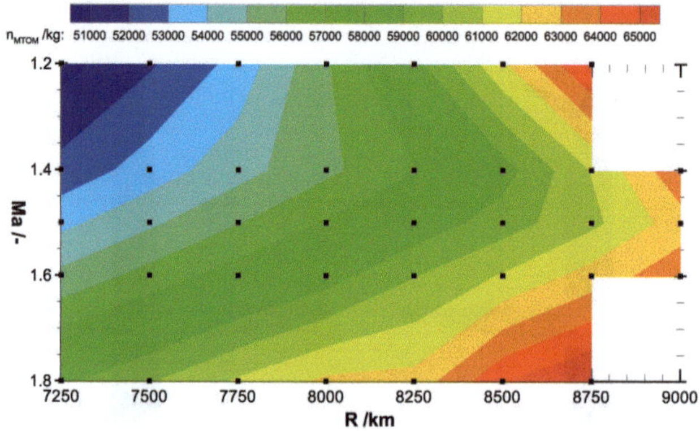


**Figure A.13:** Parameter variation results: max. take-off mass, $m_{MTOM}$

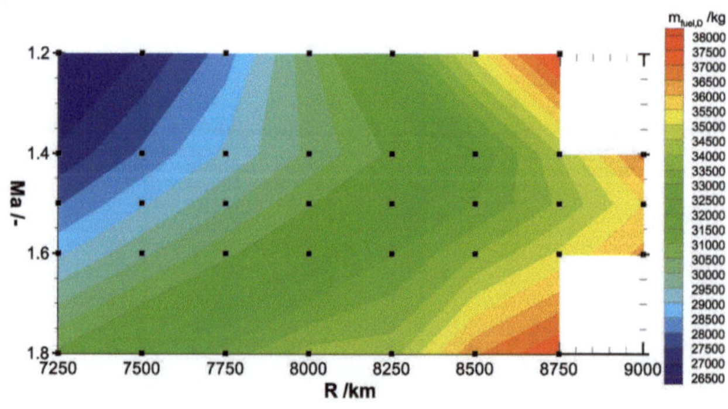


**Figure A.14:** Parameter variation results: fuel mass of the design mission, $m_{fuel,D}$

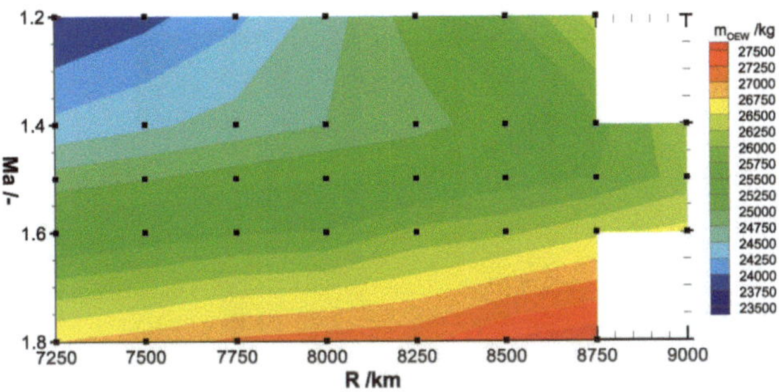


**Figure A.15:** Parameter variation results: operating empty mass, $m_{OEW}$

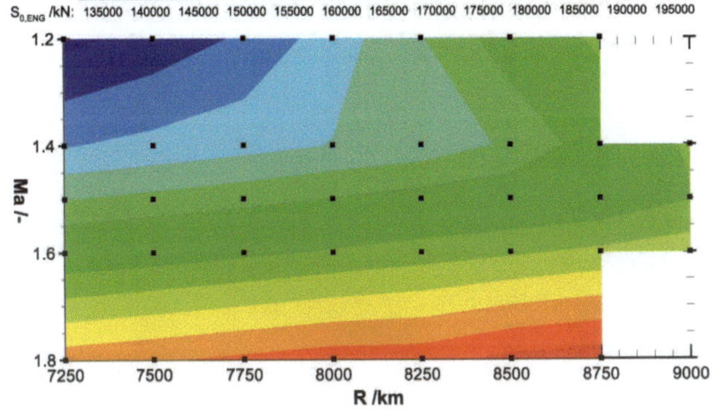


**Figure A.16:** Parameter variation results: static thrust, $S_{0,ENG}$

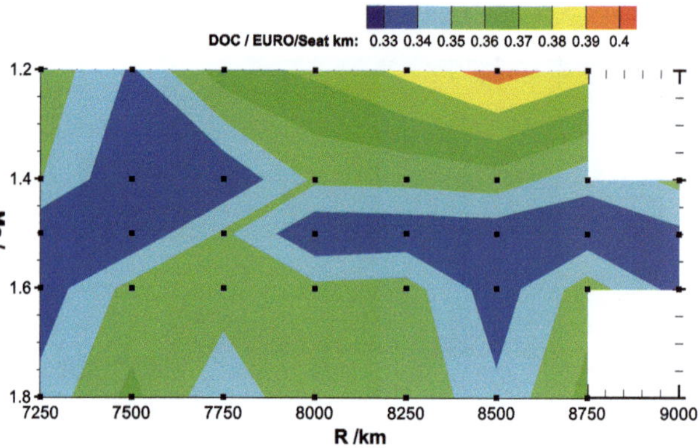


**Figure A.17:** Parameter variation results: direct operating costs (DOC)

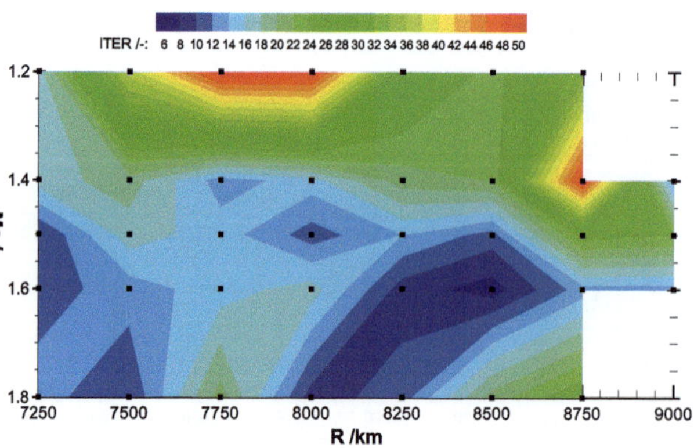


**Figure A.18:** Parameter variation results: number of design iterations

# B Tables

Table B.1: Overview of preliminary aircraft design tools

| *Acronym* | *Name* | *Origin* | *Source* |
|---|---|---|---|
| AAA | Advanced Aircraft Analysis | DARcorporation | Anemaat & Kaushik (2011) |
| ACSYNT | Aircraft Synthesis Program | NASA Ames, Virginia Tech, Phoenix | Malone & Myklebust (1996) |
| AIDA | Artificial Intelligence supported Design of Aircraft | TU Delft | La Rocca & Tooren (2007); Rentema (2004) |
| AMRaven | Adaptive Modeling Rapid Air Vehicle Engineering | US Air Force Research, Technosoft Inc. | Anemaat et al. (2007); Mukhopadhyay (2007); Stephenson et al. (2007) |
| CAPDA | Computer Aided Preliminary Design of Aircraft | TU Berlin | Xie & C. (1999) |
| CDS | Conceptual Design Shop | NASA Langley, Technosoft Inc. | Mukhopadhyay (2007) |
| CEASIOM | Computerized Environment for Aircraft Synthesis and Integrated Optimization Methods | EU FP6 Project: SimSAC | Rizzi et al. (2011); Kaenel et al. (2008) |
| FLOPS | FLight OPtimization System | NASA Langley | Nickol & A. (2009) |
| MICADO | Multidisciplinary Integrated Conceptual Aircraft Design and Optimization Environment | RWTH Aachen | Risse et al. (2012) |
| Pacelab APD | Aircraft Preliminary Design | PACE GmbH | N/A |
| PASS | Program for Aircraft Synthesis Studies | Stanford University | Antoine et al. (2005) |
| PIANO | N/A | Lissys Ltd. | Simos (2006) |
| PrADO | Preliminary Aircraft Design and Optimisation Program | TU Braunschweig | Werner-Westphal et al. (2008b) |
| PreSTo | Aircraft Preliminary Sizing Tool | Hamburg University of Applied Sciences (HAW) | Seeckt (2011) |
| RDS | Raymer's Design System | Conceptual Research Corporation | Raymer (2006, 2011) |

**Table B.2:** Default settings for SSM

| *Parameter* | *Value* | *Unit* | *Description* |
|---|---|---|---|
| CF | 0.98 | - | Calibration factor (cf. Section 3.2.4) |
| $RBEAM$ | 0.5 | - | Beam sizing factor (cf. Section 3.2.4) |
| FI | 1.0 | - | Desired failure index for all domains |
| I0MAX | 30 | - | Maximum number of iterations |
| $dFI_{max}$ | 0.15 | % | Allowed change of the maximum failure index |
| $dMi$ | 0.5 | % | Allowed change of the structural mass |
| $del(M)$ | 1.0 | kg | Arithmetic mean value of<br>change of structural mass for 3 iterations |
| $del(M)/M$ | 0.5 | % | Quotient of $del(M)$ and the arithmetic mean value<br>of structural mass for 4 iterations |

**Table B.3:** DLR-F11 wing: load cases (data from Hürlimann et al. (2012))

| *Number /-* | *Type /-* | *n /g* | *Alt /km* | *Ma /-* | *ass /kg* |
|---|---|---|---|---|---|
| 1 | Maneuver | 2.87 | 0.0 | 0.567 | 150949 |
| 2 | Landing | 3.50 | 0.0 | - | 161269 |
| 3 | Landing | 3.50 | 0.0 | - | 161269 |
| 4 | Maneuver | 2.50 | 10.0 | 0.850 | 230316 |
| 5 | Maneuver | 2.50 | 0.0 | 0.709 | 230316 |
| 6 | Maneuver | 2.50 | 6.4 | 0.850 | 150949 |
| 7 | Cruise | 1.00 | 10.0 | 0.846 | 190632 |
| 8 | Crash (horiz.) | 6.00 | 0.0 | - | 230316 |

Remark: The data has been taken directly from the source, in which load case 2 and 3 are identical.

**Table B.4:** DLR-F11 wing: minimum gage constraints

| *Group name* | *Minimum value* | *Unit* | *Material name* |
|---|---|---|---|
| Ribs | 3.5 | mm | AL 2024 |
| Spars | 4.0 | mm | AL 7050-T7451 |
| Upper Skin | 4.0 | mm | AL 7050-T7451 |
| Lower Skin | 4.0 | mm | AL 2024 |
| Upper Stringer | 300.0 | $mm^2$ | AL 7050-T7451 |
| Lower Stringer | 300.0 | $mm^2$ | AL 2024 |

**Table B.5:** DLR-F11 wing: material data

| *Material name* | *Yield strength /MPa* | *Ultimate strength /MPa* |
|---|---|---|
| AL 7050-T7451 | 469.0 | 524.0 |
| AL 2024 | 324.0 | 469.0 |

**Table B.6:** HISAC aircraft: load cases derived from V-n diagram (final design iteration)

| *LC* | *LC-type* | *Mass* | *Altitude* | *Ma* | *nz* | *Payload* | *Fuel mass* | $\Delta p$ |
|---|---|---|---|---|---|---|---|---|
| - | - | *kg* | *m* | - | - | *kg* | *kg* | $\frac{N}{m^2}$ |
| 1 | FC-BOE | 25383. | 7354. | 1.000 | 3.89 | 0. | 0. | 40441.68 |
| 2 | GC | 50103. | 0. | 0.000 | 1.40 | 0. | 24720. | 0.0 |
| 3 | FC-MOK | 50103. | 0. | 0.441 | -1.00 | 0. | 24720. | 0.0 |
| 4 | FC-BOE | 50103. | 7354. | 1.000 | 2.64 | 0. | 24720. | 40441.68 |
| 5 | GC | 50103. | 0. | 0.000 | 1.40 | 1729. | 22992. | 0.0 |
| 6 | FC-MOK | 50103. | 0. | 0.441 | -1.00 | 1729. | 22992. | 0.0 |
| 7 | FC-BOE | 50103. | 7354. | 1.000 | 2.64 | 1729. | 22992. | 40441.68 |
| 8 | LI | 36633. | 0. | 0.135 | 2.48 | 0. | 11250. | 0.0 |
| 9 | LI | 36632. | 0. | 0.135 | 2.48 | 1729. | 9521. | 0.0 |
| 10 | FC-MOK | 28320. | 17480. | 1.800 | 2.50 | 0. | 2937. | 71349.07 |
| 11 | FC-MOK | 50105. | 14341. | 1.600 | 1.00 | 728. | 23995. | 66131.68 |
| 12 | FC-BOE | 27305. | 7354. | 1.000 | 3.72 | 0. | 1924. | 40441.68 |
| 13 | FC-BOE | 27111. | 7354. | 1.000 | 3.74 | 1729. | 0. | 40441.68 |
| 14 | FC-BOE | 29035. | 7354. | 1.000 | 3.59 | 1729. | 1924. | 40441.68 |
| 15 | FC-BOE | 27111. | 6000. | 0.320 | 2.24 | 1729. | 0. | 32314.21 |

FC-BOE denotes a gust case

FC-MOK denotes a manoeuvre case

GC is a ground case

LI denotes a landing impact case

$\Delta$p is the pressure difference of cabin pressure and pressure at given altitude

**Table B.7:** Temperature analysis: geometry data of the wing

| *Parameter* | *Value* | *Unit* | *Description* |
|---|---|---|---|
| $A_W$ | 150.0 | $m^2$ | Wing area |
| $s$ | 18.46 | $m$ | Wing span |
| $x_{kink}$ | 0.4479 | - | Dimensionless coordinate of the kink |
| $\lambda_{rk}$ | 0.4048 | - | Taper ratio from root to kink |
| $\lambda_{kt}$ | 0.0988 | - | Taper ratio from kink to wing tip |
| $l_{root}$ | 16.64 | $m$ | Root chord |
| $l_{ref}$ | 11.21 | $m$ | Reference chord |
| $\delta_{root}$ | 3.0 | - | Rel. thickness to chord ratio at wing root |
| $\delta_{kink}$ | 2.5 | - | Rel. thickness to chord ratio at kink |
| $\delta_{tip}$ | 2.5 | - | Rel. thickness to chord ratio at wing tip |
| $\varphi_{LE,rk}$ | 72.6 | $deg$ | Leading edge sweep root to kink |
| $\varphi_{LE,kt}$ | 52.1 | $deg$ | Leading edge sweep kink to wing tip |
| $x_{wing}$ | 12.77 | $m$ | x-position in global coordinate system |
| $y_{wing}$ | 0.0 | $m$ | y-position in global coordinate system |
| $z_{wing}$ | -0.086 | $m$ | z-position in global coordinate system |

**Table B.8:** Temperature analysis: geometry data of the horizontal tailplane (HTP)

| *Parameter* | *Value* | *Unit* | *Description* |
|---|---|---|---|
| $A_{HTP}$ | 9.0 | $m^2$ | HTP area |
| $s_{HTP}$ | 5.612 | $m$ | HTP span |
| $\lambda_{HTP}$ | 0.257 | - | Taper ratio |
| $l_{root,HTP}$ | 2.5516 | $m$ | Root chord |
| $\delta_{root,HTP}$ | 4.5 | - | Thickness to chord ratio at HTP root |
| $\delta_{tip,HTP}$ | 4.0 | - | Thickness to chord ratio at HTP tip |
| $\varphi_{25,HTP}$ | 45.0 | $deg$ | Sweep of $\frac{l}{4}$-line |
| $x_{HTP}$ | 32.7584 | $m$ | x-position in global coordinate system |
| $y_{HTP}$ | 0.0 | $m$ | y-position in global coordinate system |
| $z_{HTP}$ | 0.66 | $m$ | z-position in global coordinate system |

**Table B.9:** Temperature analysis: geometry data of the vertical tailplane (VTP)

| *Parameter* | *Value* | *Unit* | *Description* |
|---|---|---|---|
| $A_{VTP}$ | 12.195 | $m^2$ | VTP area |
| $s_{VTP}$ | 3.98 | $m$ | VTP height |
| $\lambda_{VTP}$ | 0.3623 | - | Taper ratio |
| $l_{root,VTP}$ | 4.498 | $m$ | Root chord |
| $\delta_{root,VTP}$ | 4.5 | - | Thickness to chord ratio at VTP root |
| $\delta_{tip,VTP}$ | 4.0 | - | Thickness to chord ratio at VTP tip |
| $\varphi_{25,VTP}$ | 46.9 | $deg$ | Sweep of $\frac{l}{4}$-line |
| $x_{VTP}$ | 31.312 | $m$ | x-position in global coordinate system |
| $y_{VTP}$ | 0.0 | $m$ | y-position in global coordinate system |
| $z_{VTP}$ | 1.062 | $m$ | z-position in global coordinate system |

**Table B.10:** Basic aircraft: geometry data of the fuselage

| *Parameter* | *Value* | *Unit* | *Description* |
|---|---|---|---|
| $l_{GR}$ | 36.018 | $m$ | Length of the fuselage |
| $d_{ARZ}$ | 2.030 | $m$ | Maximum height |
| $d_{ARY}$ | 2.030 | $m$ | Maximum width |
| $V_{fuel,fus}$ | 27.200 | $m^3$ | Fuel volume |

**Table B.11:** Basic aircraft: geometry data of the wing

| *Parameter* | *Value* | *Unit* | *Description* |
|---|---|---|---|
| $A_W$ | 150.0 | $m^2$ | Wing area |
| $s$ | 18.453 | $m$ | Wing span |
| $x_{kink}$ | 0.4479 | - | Dimensionless coordinate of the kink |
| $\lambda_{rk}$ | 0.4048 | - | Taper ratio from root to kink |
| $\lambda_{kt}$ | 0.0988 | - | Taper ratio from kink to wing tip |
| $l_{root}$ | 16.678 | $m$ | Root chord |
| $l_{ref}$ | 11.21 | $m$ | Reference chord |
| $\delta_{root}$ | 3.0 | - | Rel. thickness to chord ratio at wing root |
| $\delta_{kink}$ | 2.5 | - | Rel. thickness to chord ratio at kink |
| $\delta_{tip}$ | 2.5 | - | Rel. thickness to chord ratio at wing tip |
| $\varphi_{LE,rk}$ | 72.6 | $deg$ | Leading edge sweep root to kink |
| $\varphi_{LE,kt}$ | 52.1 | $deg$ | Leading edge sweep kink to wing tip |
| $\epsilon_W$ | 1.5 | $deg$ | Angle of incidence |
| $x_{wing}$ | 12.77 | $m$ | x-position in global coordinate system |
| $y_{wing}$ | 0.0 | $m$ | y-position in global coordinate system |
| $z_{wing}$ | -0.086 | $m$ | z-position in global coordinate system |
| $V_{fuel,wing}$ | 17.500 | $m^3$ | Fuel volume |

Table B.12: Basic aircraft: geometry data of canard

| *Parameter* | *Value* | *Unit* | *Description* |
|---|---|---|---|
| $A_{Canard}$ | 7.0 | $m^2$ | Canard area |
| $s_{Canard}$ | 4.056 | $m$ | Canard span |
| $\lambda_{Canard}$ | 0.257 | - | Taper ratio |
| $l_{root,Canard}$ | 2.746 | $m$ | Root chord |
| $\delta_{root,Canard}$ | 4.5 | - | Thickness to chord ratio at canard root |
| $\delta_{tip,Canard}$ | 4.0 | - | Thickness to chord ratio at canard tip |
| $\varphi_{25,Canard}$ | 45.0 | $deg$ | Sweep of $\frac{l}{4}$-line |
| $x_{Canard}$ | 2.064 | $m$ | x-position in global coordinate system |
| $y_{Canard}$ | 0.0 | $m$ | y-position in global coordinate system |
| $z_{Canard}$ | -1.320 | $m$ | z-position in global coordinate system |

Table B.13: Basic aircraft: geometry data of the vertical tailplane (VTP)

| *Parameter* | *Value* | *Unit* | *Description* |
|---|---|---|---|
| $A_{VTP}$ | 12.195 | $m^2$ | VTP area |
| $s_{VTP}$ | 3.98 | $m$ | VTP height |
| $\lambda_{VTP}$ | 0.3623 | - | Taper ratio |
| $l_{root,VTP}$ | 4.498 | $m$ | Root chord |
| $\delta_{root,VTP}$ | 4.5 | - | Thickness to chord ratio at VTP root |
| $\delta_{tip,VTP}$ | 4.0 | - | Thickness to chord ratio at VTP tip |
| $\varphi_{25,VTP}$ | 46.9 | $deg$ | Sweep of $\frac{l}{4}$-line |
| $x_{VTP}$ | 31.313 | $m$ | x-position in global coordinate system |
| $y_{VTP}$ | 0.0 | $m$ | y-position in global coordinate system |
| $z_{VTP}$ | 1.062 | $m$ | z-position in global coordinate system |

**Table B.14:** Basic aircraft: mass break down of $m_{OEW}$

| *Parameter number /-* | *Parameter name /-* | *Parameter symbol /-* | *Value / kg* | *Value / %* |
|---|---|---|---|---|
| 1 | Wing | $m_{Wing}$ | 5230. | 20.630 |
| 2 | Fuselage | $m_{Fuselage}$ | 3243. | 12.793 |
| 3 | Canard | $m_{Canard}$ | 212. | 0.837 |
| 4 | Vertical Tailplane | $m_{VTP}$ | 253. | 0.998 |
| 5 | Winglets | $m_{WL}$ | 0. | 0.000 |
| 6 | Fairings | $m_{Fair}$ | 0. | 0.000 |
| 7 | Pylons | $m_{PYL}$ | 0. | 0.000 |
| 8 | Landing Gear | $m_{Gear}$ | 1503. | 5.927 |
| 9 | Propulsion Unit | $m_{punit}$ | 7562. | 29.826 |
| 10 | Systems | $m_{systems}$ | 5630. | 22.205 |
| 11 | Standard Items Variations | $m_{SIV}$ | 680. | 2.682 |
| | Basic Empty Mass (Weight) (1-11) | $m_{BEW}$ | 24312. | 95.899 |
| 12 | Operational Items | $m_{OI}$ | 1040. | 4.101 |
| | Operational Empty Mass (Weight) (1-12) | $m_{OEW}$ | 25352. | 100.000 |

**Table B.15:** Basic aircraft: data of the propulsion system

| *Parameter name /-* | *Parameter symbol /-* | *Value* | *Unit* |
|---|---|---|---|
| Specific fuel consumption at: | | | |
| Take-off | $SFC_{TO}$ | 0.06267 | $\frac{kg}{Nh}$ |
| Climb | $SFC_{Climb}$ | 0.07714 | $\frac{kg}{Nh}$ |
| Cruise | $SFC_{Cruise}$ | 0.08555 | $\frac{kg}{Nh}$ |
| Descent | $SFC_{D}$ | 0.07575 | $\frac{kg}{Nh}$ |
| Data for one engine: | | | |
| Static thrust | $S_{0,ENG}$ | 166.96 | $kN$ |
| Bypass ratio | $BPR$ | 0.36 | – |
| Total pressure ratio | $\pi_{total}$ | 32.4 | – |
| Allowed temperature max. At turbine entrance | $TET_{max}$ | 1755. | $K$ |
| Length | $l_{engine}$ | 4977.5 | $mm$ |
| Diameter | $D_{engine}$ | 1230.6 | $mm$ |
| Engine mass | $m_{engine}$ | 2308.4 | $kg$ |
| Electrical power extraction | $P_{elec}$ | 50.00 | $kW$ |
| Bleed air extraction (mass flow rate) | $\dot{m}_{bleed}$ | 0.09 | $\frac{kg}{s}$ |

Table B.16: Basic aircraft: load cases derived from V-n diagram (final design iteration)

| *LC* | *LC-type* | *Mass* | *Altitude* | *Ma* | *nz* | *Payload* | *Fuel mass* | $\Delta p$ |
|---|---|---|---|---|---|---|---|---|
| - | - | *kg* | *m* | - | - | *kg* | *kg* | $\frac{N}{m^2}$ |
| 1 | FC-BOE | 25350. | 6741. | 0.945 | 3.71 | 0. | 0. | 36917.48 |
| 2 | GC | 55408. | 0. | 0.000 | 1.40 | 0. | 30058. | 0.0 |
| 3 | FC-MOK | 55408. | 0. | 0.441 | -1.00 | 0. | 30058. | 0.0 |
| 4 | FC-BOE | 55408. | 0. | 0.529 | 2.50 | 0. | 30058. | 0.0 |
| 5 | GC | 55408. | 0. | 0.000 | 1.40 | 1729. | 28328. | 0.0 |
| 6 | FC-MOK | 55408. | 0. | 0.441 | -1.00 | 1729. | 28328. | 0.0 |
| 7 | FC-BOE | 55408. | 0. | 0.529 | 2.50 | 1729. | 28328. | 0.0 |
| 8 | LI | 38811. | 0. | 0.170 | 3.35 | 0. | 13460. | 0.0 |
| 9 | LI | 38811. | 0. | 0.170 | 3.35 | 1729. | 11731. | 0.0 |
| 10 | FC-MOK | 28085. | 18133. | 1.800 | 2.50 | 0. | 2735. | 72145.88 |
| 11 | FC-MOK | 55410. | 14341. | 1.600 | 1.00 | 728. | 29331. | 66131.68 |
| 12 | FC-BOE | 27017. | 6741. | 0.945 | 3.57 | 0. | 1666. | 36917.48 |
| 13 | FC-BOE | 27080. | 6741. | 0.945 | 3.57 | 1729. | 0. | 36917.48 |
| 14 | FC-BOE | 28746. | 6741. | 0.945 | 3.45 | 1729. | 1666. | 36917.48 |
| 15 | FC-BOE | 27080. | 6000. | 0.403 | 2.29 | 1729. | 0. | 32314.00 |

FC-BOE denotes a gust case

FC-MOK denotes a manoeuvre case

GC is a ground case

LI denotes a landing impact case

$\Delta$p is the pressure difference of cabin pressure and pressure at given altitude

**Table B.17:** Basic aircraft: mission data results

| | | Mission | | | |
|---|---|---|---|---|---|
| *Parameter* | *Unit* | *Design* | *Max. payload* | *Max. fuel* | *Ferry flight* |
| Range | $km$ | 7408. | 7000. | 7747. | 7747. |
| Flight time | $h$ | 4.96 | 4.72 | 5.16 | 5.16 |
| Nb. passengers | - | 8 | 19 | 0 | 0 |
| Cargo | $kg$ | 0. | 0. | 0. | 0. |
| Payload | $kg$ | 728. | 1729. | 0. | 0. |
| Fuel mass | $kg$ | 29331. | 28328. | 30058. | 30058. |
| Cruise Mach number | - | 1.60 | 1.60 | 1.60 | 1.60 |
| Cruise speed | $\frac{km}{h}$ | 1699.60 | 1699.60 | 1699.60 | 1699.60 |
| Altitude at: | | | | | |
| begin of cruise | $km$ | 14.34 | 14.34 | 14.34 | 14.34 |
| end of cruise | $km$ | 17.95 | 17.72 | 18.13 | 18.13 |
| Lift-to-drag ratio: | | | | | |
| begin of cruise | - | 5.59 | 5.59 | 5.60 | 5.60 |

**Table B.18:** Parameter variation: results for range and $Ma$ combinations

(a) Design range: 7250 km

| *Parameter* | *Unit* | | | | | |
|---|---|---|---|---|---|---|
| $Ma$ | $/-$ | 1.2 | 1.4 | 1.5 | 1.6 | 1.8 |
| $R_4$ | $/km$ | | | 7250 | | |
| $m_{MTOM}$ | $kg$ | 50630. | 52324. | 53725. | 55253. | 58762. |
| $m_{OEW}$ | $kg$ | 23725. | 24313. | 24868. | 25446. | 26751. |
| $m_{fuel,max}$ | $kg$ | 26905. | 28015. | 28857. | 29811. | 32014. |
| $W_{S,max}$ | $\frac{kg}{m^2}$ | 337.52 | 348.84 | 358.15 | 368.37 | 391.77 |
| $S_0$ | $kN$ | 269.26 | 291.72 | 312.20 | 333.50 | 377.92 |
| $ALT_{begin,cruise,D}$ | $km$ | | | 14.341 | | |
| $ALT_{end,cruise,D}$ | $km$ | 17.848 | 17.860 | 17.861 | 17.872 | 17.913 |
| $L/D_{begin,cruise,D}$ | $m$ | 7.244 | 6.30 | 5.93 | 5.58 | 4.97 |
| $V_{AT}$ | $\frac{km}{h}$ | 233. | 237. | 240. | 243. | 250. |
| | $kts$ | 126. | 128. | 129. | 131. | 135. |
| $l_{RWY,T}$ | $m$ | 2342. | 2389. | 2428. | 2476. | 2606. |
| $l_{RWY,LD}$ | $m$ | 1305. | 1369. | 1425 | 1484. | 1611. |
| DOC | $\frac{EURO}{SKM}$ | 0.35721 | 0.35017 | 0.33158 | 0.33394 | 0.34317 |

(b) Design range: 7500 km

| *Parameter* | *Unit* | | | | | |
|---|---|---|---|---|---|---|
| $Ma$ | $/-$ | 1.2 | 1.4 | 1.5 | 1.6 | 1.8 |
| $R_4$ | $/km$ | | | 7500 | | |
| $m_{MTOM}$ | $kg$ | 51868. | 53461. | 54816. | 56374. | 59912. |
| $m_{OEW}$ | $kg$ | 23921. | 24454. | 24988. | 25581. | 26878. |
| $m_{fuel,max}$ | $kg$ | 27947. | 29003. | 29825. | 30797. | 33032. |
| $W_{S,max}$ | $\frac{kg}{m^2}$ | 345.80 | 356.39 | 365.42 | 375.83 | 399.41 |
| $S_0$ | $kN$ | 275.88 | 295.07 | 315.34 | 336.57 | 380.88 |
| $ALT_{begin,cruise,D}$ | $km$ | | | 14.341 | | |
| $ALT_{end,cruise,D}$ | $km$ | 17.949 | 17.959 | 17.961 | 17.969 | 18.008 |
| $L/D_{begin,cruise,D}$ | $m$ | 7.31 | 6.37 | 5.99 | 5.64 | 5.03 |
| $V_{AT}$ | $\frac{km}{h}$ | 236. | 239. | 242. | 245. | 252. |
| | $kts$ | 127. | 129. | 130. | 132. | 136. |
| $l_{RWY,T}$ | $m$ | 2407. | 2457. | 2493. | 2543. | 2674. |
| $l_{RWY,LD}$ | $m$ | 1340. | 1399. | 1452. | 1512. | 1638. |
| DOC | $\frac{EURO}{SKM}$ | 0.33861 | 0.33107 | 0.33162 | 0.35333 | 0.36264 |

**Table B.18:** continued

(c) Design range: 7750 km

| *Parameter* | *Unit* | | | | | |
|---|---|---|---|---|---|---|
| $Ma$ | $/-$ | 1.2 | 1.4 | 1.5 | 1.6 | 1.8 |
| $R_4$ | $/km$ | | | 7750 | | |
| $m_{MTOM}$ | $kg$ | 53350. | 54621. | 55957. | 57505. | 61145. |
| $m_{OEW}$ | $kg$ | 24190. | 24603. | 25132. | 25724. | 27051. |
| $m_{fuel,max}$ | $kg$ | 29129. | 30018. | 30828. | 31783. | 34096. |
| $W_{S,max}$ | $\frac{kg}{m^2}$ | 355.49 | 364.15 | 372.05 | 383.37 | 407.66 |
| $S_0$ | $kN$ | 283.54 | 298.76 | 318.77 | 339.83 | 384.50 |
| $ALT_{begin,cruise,D}$ | $km$ | | | 14.341 | | |
| $ALT_{end,cruise,D}$ | $km$ | 18.064 | 18.056 | 18.057 | 18.065 | 18.100 |
| $L/D_{begin,cruise,D}$ | $m$ | 7.32 | 6.44 | 6.05 | 5.70 | 5.00 |
| $V_{AT}$ | $\frac{km}{h}$ | 238. | 241. | 243. | 246. | 253. |
| | $kts$ | 129. | 130. | 131. | 133. | 137. |
| $l_{RWY,T}$ | $m$ | 2487. | 2527. | 2560. | 2611. | 2745. |
| $l_{RWY,LD}$ | $m$ | 1383. | 1429. | 1481. | 1539. | 1668. |
| DOC | $\frac{EURO}{SKM}$ | 0.36623 | 0.33086 | 0.35181 | 0.35502 | 0.34196 |

(d) Design range: 8000 km

| *Parameter* | *Unit* | | | | | |
|---|---|---|---|---|---|---|
| $Ma$ | $/-$ | 1.2 | 1.4 | 1.5 | 1.6 | 1.8 |
| $R_4$ | $/km$ | | | 8000 | | |
| $m_{MTOM}$ | $kg$ | 55903. | 55774. | 57109. | 58518. | 62343. |
| $m_{OEW}$ | $kg$ | 24625. | 24754. | 25262. | 25768. | 27165. |
| $m_{fuel,max}$ | $kg$ | 31274. | 31023. | 31847. | 32752. | 35175. |
| $W_{S,max}$ | $\frac{kg}{m^2}$ | 372.78 | 371.84 | 380.75 | 390.12 | 415.61 |
| $S_0$ | $kN$ | 297.20 | 302.68 | 322.76 | 343.39 | 388.03 |
| $ALT_{begin,cruise,D}$ | $km$ | | | 14.341 | | |
| $ALT_{end,cruise,D}$ | $km$ | 18.245 | 18.154 | 18.151 | 18.164 | 18.192 |
| $L/D_{begin,cruise,D}$ | $m$ | 7.23 | 6.50 | 6.12 | 5.76 | 5.16 |
| $V_{AT}$ | $\frac{km}{h}$ | 243. | 243. | 245. | 248. | 255. |
| | $kts$ | 131. | 131. | 132. | 134. | 142. |
| $l_{RWY,T}$ | $m$ | 2622. | 2596. | 2628. | 2668. | 2814. |
| $l_{RWY,LD}$ | $m$ | 1455. | 1458. | 1510. | 1562. | 1695. |
| DOC | $\frac{EURO}{SKM}$ | 0.37190 | 0.35162 | 0.33129 | 0.35302 | 0.35914 |

**Table B.18:** continued

(e) Design range: 8250 km

| *Parameter* | *Unit* | | | | | |
|---|---|---|---|---|---|---|
| $Ma$ | $/-$ | 1.2 | 1.4 | 1.5 | 1.6 | 1.8 |
| $R_4$ | $/km$ | | | 8250 | | |
| $m_{MTOM}$ | $kg$ | 58659. | 56899. | 58304. | 59949. | 63586. |
| $m_{OEW}$ | $kg$ | 33528. | 24874. | 25428. | 26077. | 27381. |
| $m_{fuel,max}$ | $kg$ | 25128. | 32029. | 32879. | 33871. | 36134. |
| $W_{S,max}$ | $\frac{kg}{m^2}$ | 391.04 | 379.35 | 388.71 | 399.65 | 423.90 |
| $S_0$ | $kN$ | 311.62 | 306.13 | 326.30 | 347.36 | 391.28 |
| $ALT_{begin,cruise,D}$ | $km$ | | | 14.341 | | |
| $ALT_{end,cruise,D}$ | $km$ | 18.426 | 18.251 | 18.244 | 18.246 | 18.280 |
| $L/D_{begin,cruise,D}$ | $m$ | 7.15 | 6.56 | 6.18 | 5.84 | 5.22 |
| $V_{AT}$ | $\frac{km}{h}$ | 247. | 244. | 247. | 250. | 257. |
| | $kts$ | 134. | 132. | 133. | 135. | 138. |
| $l_{RWY,T}$ | $m$ | 2772. | 2665. | 2700. | 2754. | 2890. |
| $l_{RWY,LD}$ | $m$ | 1531. | 1487. | 1540. | 1600. | 1727. |
| DOC | $\frac{EURO}{SKM}$ | 0.38259 | 0.35266 | 0.33277 | 0.35433 | 0.36306 |

(f) Design range: 8500 km

| *Parameter* | *Unit* | | | | | |
|---|---|---|---|---|---|---|
| $Ma$ | $/-$ | 1.2 | 1.4 | 1.5 | 1.6 | 1.8 |
| $R_4$ | $/km$ | | | 8500 | | |
| $m_{MTOM}$ | $kg$ | 61808. | 58541. | 59521. | 61099. | 64879. |
| $m_{OEW}$ | $kg$ | 25726. | 25175. | 25593. | 26191. | 27522. |
| $m_{fuel,max}$ | $kg$ | 36085. | 33364. | 33926. | 34905. | 36447. |
| $W_{S,max}$ | $\frac{kg}{m^2}$ | 412.07 | 390.25 | 396.82 | 405.30 | 394.76 |
| $S_0$ | $kN$ | 328.25 | 313.77 | 329.69 | 350.48 | 432.54 |
| $ALT_{begin,cruise,D}$ | $km$ | | | 14.341 | | |
| $ALT_{end,cruise,D}$ | $km$ | 18.612 | 18.360 | 18.335 | 18.338 | 18.365 |
| $L/D_{begin,cruise,D}$ | $m$ | 7.06 | 6.58 | 6.25 | 5.90 | 5.29 |
| $V_{AT}$ | $\frac{km}{h}$ | 252. | 247. | 249. | 252. | 259. |
| | $kts$ | 136. | 133. | 135. | 236. | 144. |
| $l_{RWY,T}$ | $m$ | 2944. | 2756. | 2776. | 2824. | 2969. |
| $l_{RWY,LD}$ | $m$ | 1617. | 1531. | 1570. | 1627. | 1759. |
| DOC | $\frac{EURO}{SKM}$ | 0.39498 | 0.35533 | 0.33301 | 0.33439 | 0.34206 |

**Table B.18:** continued

(g) Design range: 8750 km

| *Parameter* | *Unit* | | | | | |
|---|---|---|---|---|---|---|
| $Ma$ | $/-$ | 1.2 | 1.4 | 1.5 | 1.6 | 1.8 |
| $R_4$ | $/km$ | | | 8750 | | |
| $m_{MTOM}$ | $kg$ | 65129. | 61052. | 60755. | 62444. | 66125. |
| $m_{OEW}$ | $kg$ | 26345. | 25652. | 25761. | 26429. | 27695. |
| $m_{fuel,max}$ | $kg$ | 37874. | 35412. | 34996. | 36012. | 37520. |
| $W_{S,max}$ | $\frac{kg}{m^2}$ | 434.31 | 407.13 | 405.05 | 416.27 | 440.80 |
| $S_0$ | $kN$ | 345.64 | 328.27 | 333.26 | 354.28 | 397.88 |
| $ALT_{begin,cruise,D}$ | $km$ | | | 14.341 | | |
| $ALT_{end,cruise,D}$ | $km$ | 18.793 | 18.509 | 18.424 | 18.421 | 18.451 |
| $L/D_{begin,cruise,D}$ | $m$ | 6.99 | 6.55 | 6.31 | 5.96 | 5.35 |
| $V_{AT}$ | $\frac{km}{h}$ | 258. | 251. | 251. | 254. | 261. |
| | $kts$ | 139. | 137. | 136. | 137. | 141. |
| $l_{RWY,T}$ | $m$ | 3124. | 2890. | 2853. | 2908. | 3042. |
| $l_{RWY,LD}$ | $m$ | 1708. | 1600. | 1600. | 1662. | 1790. |
| DOC | $\frac{\text{EURO}}{SKM}$ | 0.38458 | 0.34235 | 0.33390 | 0.35543 | 0.36344 |

(h) Design range: 9000 km

| *Parameter* | *Unit* | | | |
|---|---|---|---|---|
| $Ma$ | $/-$ | 1.4 | 1.5 | 1.6 |
| $R_4$ | $/km$ | | 9000 | |
| $m_{MTOM}$ | $kg$ | 63818. | 62511. | 63646. |
| $m_{OEW}$ | $kg$ | 26231. | 26100. | 26569. |
| $m_{fuel,max}$ | $kg$ | 36677. | 36131. | 36167. |
| $W_{S,max}$ | $\frac{kg}{m^2}$ | 425.45 | 416.73 | 424.29 |
| $S_0$ | $kN$ | 344.24 | 341.42 | 358.10 |
| $ALT_{begin,cruise,D}$ | $km$ | | 14.341 | |
| $ALT_{end,cruise,D}$ | $km$ | 18.656 | 18.531 | 18.509 |
| $L/D_{begin,cruise,D}$ | $m$ | 6.52 | 6.33 | 6.02 |
| $V_{AT}$ | $\frac{km}{h}$ | 256. | 254. | 256. |
| | $kts$ | 138. | 137. | 138. |
| $l_{RWY,T}$ | $m$ | 3038. | 2952. | 2982. |
| $l_{RWY,LD}$ | $m$ | 1678. | 1648. | 1691. |
| DOC | $\frac{\text{EURO}}{SKM}$ | 0.35239 | 0.33758 | 0.33606 |

**Table B.19:** Parameter variation: payload increase from eight to ten passengers for $Ma$=1.6 and range 7408$km$

| *Parameter* | *Value* | *Unit* | *Remark* |
|---|---|---|---|
| $m_{MTOM}$ | 55951. | $kg$ | max. take-off mass |
| $m_{OEW}$ | 25524. | $kg$ | Operational Empty Mass (Weight) |
| $m_{fuel,max}$ | 30424. | $kg$ | max. fuel mass |
| $W_{S,max}$ | 373.01 | $\frac{kg}{m^2}$ | max. wing loading |
| $S_0$ | 335.40 | $kN$ | total static thrust |
| $ALT_{begin,cruise,D}$ | 14.341 | $km$ | altitude at the begin of cruise at the design point |
| $ALT_{end,cruise,D}$ | 17.934 | $km$ | altitude at the end of cruise at the design point |
| $L/D_{begin,cruise,D}$ | 5.62 | $m$ | lift-to-drag ratio at the begin of cruise at the design point |
| $V_{AT}$ | 244. , (132.) | $\frac{km}{h}$, $(kts)$ | speed at threshold |
| $l_{RWY,T}$ | 2518. | $m$ | runway length at take-off |
| $l_{RWY,LD}$ | 1501. | $m$ | runway length at landing |
| DOC | 0.33391 | $\frac{\text{EURO}}{SKM}$ | Direct Operating Costs |

**Table B.20:** Wing loading of aircraft (Data from Kopenhagen (1996), Kreuzer (2002) and Rieke (2013))

| *Aircraft* | *MTOM /kg* | *Wing area /$m^2$* | *Wing loading /$\frac{kg}{m^2}$* |
|---|---|---|---|
| Boeing 787-8 | 227900.0 | 369.1 | 617.45 |
| Concorde | 185000.0 | 358.3 | 516.33 |
| Boeing 737-800 | 78000.0 | 125.5 | 621.51 |
| Gulfstream V | 41051.0 | 92.0 | 446.21 |
| Panavia Tornado | 24500.0 | 30.0 | 816.67 |
| McDonnell Douglas F-18 | 20000.0 | 37.2 | 537.63 |

**Table B.21:** Basic aircraft: material data

(a) Basic values

| MAT ID | *Material name* | $E1$ | $E2$ | $G12$ | $\nu_{12}$ | $\rho$ |
|---|---|---|---|---|---|---|
| - | - | MPa | MPa | MPa | - | $\frac{kg}{m^3}$ |
| 1 | Alu 2024 T3 | 73800. | 73800. | 27600. | 0.33 | 2854. |
| 5 | Alu 7075 T6 | 73800. | 73800. | 27600. | 0.33 | 2854. |
| 6 | Alu 2090 T83 | 79281. | 79281. | 29644. | 0.34 | 2508. |
| 12 | CFK HTS40 12k/RTM6 | 148000. | 10000. | 4600. | 0.35 | 1518. |

(b) Strength values (stress)

| MAT ID | *Material name* | $X_t$ | $X_c$ | $Y_t$ | $Y_c$ | $S$ |
|---|---|---|---|---|---|---|
| - | - | MPa | MPa | MPa | MPa | MPa |
| 1 | Alu 2024 T3 | 430. | 430. | 430. | 430. | 262. |
| 5 | Alu 7075 T6 | 517. | 517. | 517. | 517. | 303. |
| 6 | Alu 2090 T83 | 531. | 531. | 531. | 531. | 255. |
| 12 | CFK HTS40 12k/RTM6 | 2400. | 1450. | 55.0 | 220.0 | 91. |

(c) Laminates

| Laminate ID | Stacking sequence | Material |
|---|---|---|
| 21 | $[0/45/90/-45/0]_S$ | 12 |

(d) Symbols

| Symbol | Definition |
|---|---|
| $E1$ | Modulus of elasticity in longitudinal direction,<br>also defined as the fibre direction or 1-direction |
| $E2$ | Modulus of elasticity in lateral direction,<br>also defined as the matrix direction or 2-direction |
| $G12$ | In-plane shear modulus |
| $\nu_{12}$ | Poisson's ratio |
| $\rho$ | Density |
| $X_t$ | Longitudinal tensile strength (stress) |
| $X_c$ | Longitudinal compressive strength (stress) |
| $Y_t$ | Lateral tensile strength (stress) |
| $Y_c$ | Lateral compressive strength (stress) |
| $S$ | Shear strength in the 1-2 plane (stress) |

**Table B.22:** Basic aircraft: material selection and knock-down factors for components

| *Component* | *Knock-down factor* | *Material* |
|---|---|---|
| Wing skin top | .46 | 21 |
| Wing skin bottom | .46 | 21 |
| Wing ribs | .40 | 21 |
| Wing spar web | .40 | 21 |
| Wing spars caps | .40 | 5 |
| Wing spar reinforcements | .60 | 5 |
| Canard skin top | .46 | 21 |
| Canard skin bottom | .46 | 21 |
| Canard ribs | .40 | 21 |
| Canard spar web | .40 | 21 |
| Canard spars caps | .40 | 5 |
| Canard spar reinforcements | .60 | 5 |
| VTP skin | .5 | 21 |
| VTP ribs | .50 | 21 |
| VTP spar web | .50 | 21 |
| VTP spars caps | .50 | 5 |
| VTP spar reinforcements | .50 | 5 |
| Fuselage skin (pressurized) | .3 | 21 |
| Fuselage skin (not pressurized) | .3 | 21 |
| Fuselage frame | .3 | 5 |
| Fuselage frame (wing connection) | .3 | 5 |
| Fuselage frame (canard connection) | .3 | 5 |
| Fuselage frame (VTP connection) | .3 | 5 |
| Fuselage frame (pressure bulkhead) | .3 | 1 |
| Floor beam | .35 | 5 |

**Table B.23:** Data of Jet A-1 fuel.
Data from Anon (2010) (mean values).

| *Parameter* | *Value* | *Unit* |
|---|---|---|
| Density | 808.8 | $\frac{kg}{m^3}$ |
| Calorific value | 43147.0 | $\frac{kJ}{kg}$ |
| Price (PrADO, January 2014) | 0.55 | $\frac{\text{EURO}}{kg}$ |

# C Methods and Formulae

This Appendix provides additional information on methods and formulae that did not find space in the main chapters of this thesis, but are worth of being presented.

## C.1 Estimation of Viscous Drag

The estimation of viscous drag as used in this thesis is summarized here. The formulae presented are taken from Heinze (2005) who based his methodology on Hoerner (1965) and Fiecke (1958). Howe (2000) and Raymer (2006) describe similar approaches. Heinze (2005) proposes the following four steps to be undertaken to estimate zero drag, of which friction drag is a part.

1. Split the aircraft into $n$ components
2. Determine drag coefficients for each component (reference area for the i-th component is its wetted surface.)
   a) Determine the friction drag coefficient $C_{Fi}$
   b) Account for pressure drag by applying a form factor $FF_i$
   c) Account for compressibility effects
3. Calculate the total zero drag coefficient
4. Account for interference drag
   a) Statistical correction of zero drag
   b) Addition of interference drag for components

In the context of this work, steps 2b, 2c and 4 are irrelevant, since they are not related to viscous drag. Heinze (2005) defines four sets of equations for the estimation of viscous drag. The appropriate set to be applied depends on the Reynolds number. In this case, the Reynolds number, $Re$, is calculated for a given component and its respective reference length and compared to a range of values $Re_1$ to $Re_3$ as shown in the equations C.1 to C.3, where $k_S$ denotes the equivalent sand roughness height and $l$ the reference length of the component.

$$Re_1 = \frac{100}{k_S/l} \tag{C.1}$$

$$Re_2 = \frac{10^{2.086}}{(k_S/l)^{1.0652}} \tag{C.2}$$

$$Re_3 = \frac{10^{2.88}}{(k_S/l)^{1.074}} \tag{C.3}$$

The relevant skin friction coefficient $C_{Fi}$ is determined by equations C.4 to C.7.
Section 1: $Re \leq Re_1$

$$C_{Fi} = C_{F1} = \frac{0.455}{(\lg Re_1)^{2.58}} \tag{C.4}$$

Section 2: $Re_1 < Re \leq Re_2$

$$C_{Fi} = C_{F12} = C_{F1\div 2} = 10^{\left[\lg C_{F1} + \frac{\lg C_{F2} - \lg C_{F1}}{\lg Re_2 - \lg Re_1} \cdot (\lg Re - \lg Re_1)\right]} \tag{C.5}$$

Section 3: $Re_2 < Re \leq Re_3$

$$C_{Fi} = C_{F23} = C_{F2\div 3} = 10^{\left[\lg C_{F2} + \frac{\lg C_{F3} - \lg C_{F2}}{\lg Re_3 - \lg Re_2} \cdot (\lg Re - \lg Re_2)\right]} \tag{C.6}$$

Section 4: $Re_3 < Re$

$$C_{Fi} = C_{F3} = [1.89 + 1.62 \cdot \lg (l/k_S)]^{-2.5} \tag{C.7}$$

Equation C.8 is needed to complete the set of equations.

$$C_{F2} = [0.8606 - 0.01145 \cdot \lg (k_S/l)] \cdot C_{F3} \tag{C.8}$$

The skin friction coefficients are finally used to determine the coefficient for viscous drag as shown in equation C.9, where $A_{ref}$ denotes the reference area and $A_{comp}$ the wetted surface of the component.

$$C_{drag,v} = \frac{\sum_{i=1}^{n} C_{Fi} \cdot A_i}{A_{ref}} \tag{C.9}$$

The result contributes to overall aircraft drag.

## C.2 Determination of Stall Speed and Speed at Threshold

The calculation of the speed at threshold, $V_{AT}$, is defined in FAR 25 and corresponding European legislation as shown in equation C.10, where $V_{S0}$ is the stall speed.

$$V_{AT} = 1.3 \cdot V_{S0} \tag{C.10}$$

The stall speed is calculated in equation C.11. This is the formula for the determination of the lift coefficient, simply rearranged and solved for the velocity. Since the interest is to determine the stall speed at landing, the values are those for the aircraft in landing configuration. Here $m_{max,L}$ is the maximum landing mass, $A_W$

is the wing area and $C_{lift,max,L}$ represents the maximum lift coefficient at landing. The letter $g$ is the known gravitational acceleration.

$$V_{S0}^2 = \frac{2 \cdot g \cdot \frac{m_{max,L}}{A_W}}{\rho \cdot C_{lift,max,L}} \tag{C.11}$$

ICAO categorizes aircraft according to their speed at threshold (Anon, 2006, Section 4, Chapter 1.3). The mentioned HISAC aircraft for example has been categorized in category C with $121kts \leq V_{AT} < 141kts$.

## C.3 Wing Loading

The wing loading $W_S$ is defined as the ratio between the weight of an aircraft, $G$, and the area of the wing, $A_W$. The relation is given in equation C.12, where the letter $g$ is the known gravitational acceleration and $m$ the mass of the aircraft. In steady flight and at take-off, the lift $L$ equals weight as long as the aircraft is not in climb nor in descent. Under this assumption, equation C.12 can further be developed to equation C.13 in which lift $L$ has been replaced by the lift coefficient $C_{lift}$, introducing $V$ as the velocity and $\rho$ as the density.

$$W_S = \frac{m \cdot g}{A_W} = \frac{G}{A_W} \tag{C.12}$$

$$W_S = \frac{L}{A_W} = \frac{C_{lift} \cdot \rho \cdot V^2}{2} \tag{C.13}$$

Solving equation C.13 for the velocity $V$ results in equation C.14.

$$V = \sqrt{\frac{2 \cdot W_S}{C_{lift} \cdot \rho}} \tag{C.14}$$

The first obvious result of this manipulation is that a low wing loading will require lower speeds. This results in the need of shorter runways for take-off and landing[1]. In general, a reduction of the wing loading will result in a larger wing, which will most likely produce higher drag under cruise conditions[2]. At this stage, two opposing trends can be seen. A high wing loading will give better cruise efficiency while a low wing loading will influence take-off and landing performance positively. For more information on this matter, see, e.g., Heinze (2007); Raymer (2006); Roskam (1991).

[1]For take-off and landing, $C_L$ has to be replaced by a value that is appropriate for the condition.

[2]Additionally, the wing area directly influences the range calculations, as can be seen, e.g. in the Breguet Range Equation, which will not be detailed here.